POOF!

Plant Profits in Peril

A Patent & Trademark Storybook for Horticulturists

Allison Frederick

Published 2014.
Printed in the United States of America.

ISBN 978-0-9915646-3-7

Library of Congress Control Number 2014908402

Published by BookCrafters,
Parker, Colorado.
http://bookcrafters.net
bookcrafters@comcast.net

This book may be ordered from
www.bookcrafters.net
and other online bookstores.

This book is dedicated to
all of my talented clients.
They provide me the opportunity
to enjoy one's elusive dream
of "loving the work you do."

A portion of the proceeds from your purchase
of this book supports the tireless labors
of honeybee pollinators, without whom,
we wouldn't have plants to patent!

For more details, please visit:
http://www.greenhouseips.com/honeybee-donation.html

The honeybee gift program is managed by the not-for-profit organization, Heifer International.

Table of Contents

ACKNOWLEDGEMENTS

I began my career working in the pediatric oncology unit of the Denver Children's Hospital. When you are constantly surrounded by terminally ill children, you realize that any moment, and any seemingly benign situation, can suddenly become precarious, sending a young one into an intensive care unit in a matter of minutes. When I transitioned to the intellectual property field, an attorney helped me understand how my aptitude for attentiveness and vigilance applied equally to business settings, and was particularly valuable to the world of intellectual property. I wish to thank patent attorney, Darin Gibby, who helped me recognize this connection. His dedication to offering quality client service has inspired the present model for my company, GRIPS, LLC.

With much appreciation, I also wish to thank patent attorney, Bill Vobach, who is, arguably, the most humble and most conscientious attorney I've ever had the pleasure of working with. I rely on his expertise for creating powerful solutions for challenging legal problems. Bill's passion for understanding patent law reveals pathways and opportunities for his clients that others thought improbable. I am grateful for all his support over the years and for making an effort to truly understand the values and objectives of my business, my clients, and those in the horticulture industry.

Last, but not least, I am indebted and grateful to Sean Hannon. Without his editorial commitment, the concepts and stories in this book would still be racing through my mind where they have been churning for at least seven years! This project is just one of the numerous ways he makes my life better on a daily basis.

FOREWORD

Protecting the intellectual property (IP) of a plant related business is a unique endeavor. Even among IP professionals, protecting plants is an area in which only a select few specialize. In other books, the subject of IP protection for plants has been relegated to short end chapters that seem like a mere after-thought to other topics. Indeed, most books are devoted to protecting electrical, mechanical, and biomedical subject matter, rather than plants.

Allison Frederick provides a great service to the plant breeding industry by authoring a book that actually focuses on IP protection for plants and plant related businesses. Moreover, she directs the book toward a unique audience – plant breeders. While this book will be useful to IP professionals and those with an interest in protecting plants, plant breeders in particular will find that the book addresses the issues that are most important to protecting the intellectual capital of their businesses.

POOF! is filled with real-world examples and is written in an entertaining and easy to read style. The various fact patterns used to teach the lessons of each chapter draw on real world situations and actual lawsuits. Most importantly, however, the examples have been carefully selected to teach plant professionals how to plan ahead so as to avoid the pitfalls and traps that can cause their potential profits to go POOF!

As noted in the book, most IP professionals who focus on the plant breeding industry do not specialize in all areas of intellectual property. Similarly, most books to not treat all areas of IP protection for plants. Therefore, one must typically read about trademark protection for plants in a book strictly devoted to trademarks, patent protection for plants in a book strictly devoted to patents, plant variety protection for plants in a book strictly devoted to the PVPA, etc. Allison brings these subjects together in one resource – providing a comprehensive picture for the plant professional who wants to protect his/her business.

One will appreciate from reading POOF! that it is written with an understanding of the perspective of plant professionals. Having worked with plant related businesses for a long time, Allison understands the plant breeding industry, its customs, and its practices. Some of these long-standing practices create problems for plant breeders. Allison identifies and addresses these issues to help new and old plant breeders avoid similar problems down the road.

As with any competitive industry, one core value of

plant professionals is to save money spent on legal costs. Having known Allison for a long time, I can attest to the fact that helping business owners save money on legal costs is a keystone of her business. As you read POOF!, you will appreciate this fact through the many tips that Allison provides to achieve that goal.

Bill Vobach*
June 2014

*Bill Vobach is a registered patent attorney in Denver, Colorado. He has been practicing law since 1994 and holds a J.D. from the University of Denver College of Law. Mr. Vobach also has a B.A. from Oberlin College and a B.S.E.E. from Washington University in St. Louis.

INTRODUCTION

We've all seen live magicians make something like a rabbit, a bird, or sometimes even a car seemingly vanish into thin air with a sudden and loud "flash," "bang," or *"POOF!"* The audience gasps in astonishment and roars with applause as the magician then proceeds to make the object reappear just as effortlessly as before. In the world of magic, it is very easy to make something reappear. In the world of intellectual property, this is not so easy. *POOF!* is the perfect metaphor for this book, where people and companies, by not paying attention or being distracted by more "important" or more "interesting" things, make their valuable intellectual property assets disappear… forever.

POOF! is not your typical patent and trademark book that will tell you the precise filing procedures for patent and trademark applications. I wouldn't want to torture you to read such a book, and I wouldn't want to torture myself to write it! Instead, this is a book providing real world examples that are sometimes frightening stories of how companies and plant breeders who started off trying to benefit from intellectual property, inadvertently diluted their momentum and handed their market-place advantage on a silver platter to their competitors and their financial war chests to their attorneys.

Earlier in my career, I managed an international portfolio of over 1,000 patents for a Fortune 500 company that had a relatively large budget, but even they needed to figure out how to best leverage their patents, particularly in the international market. The same types of decisions are faced by board members, vice presidents of intellectual property divisions, directors of not-for-profits, and individual breeders. None of us can really afford the adage, "There is never time to do it right the first time… but always time to do it *again!*" To stay competitive, it is imperative that you and your team all try to do it "right" the first time.

To prepare for this book, I asked myself questions like: What do my clients really need to understand about how patents and trademarks function? If plant breeders understood some fundamental principles to guide their decision-making, what would those principles be? Who has done a great job managing their intellectual property? And, who has gotten themselves into trouble?

The principles in this book will guide you in designing your plant breeding programs in any part of the world so you keep your options open for as long as possible. Understanding these concepts can help you appreciate the potential value of good contracts and most importantly, if you already have some intellectual property, you will learn how to maintain it so you don't experience a sudden setback or complete vanishing of your intellectual property portfolio.

After reading this book you will better understand many of the different profitable opportunities that companies inadvertently give away. You'll recognize how to avoid paying exorbitant attorney fees chasing risky applications and ending up with nothing to show for it. This book can help you save money on your legal consultations because you will be prepared to ask more poignant questions of professionals who charge by the hour.

You will learn the value of taking the time to develop a solid, but not overwhelming, intellectual property strategy *before* you get too busy turning your products into gold with your Midas touch.

Legal Lingo and Other Ho-Hums

For the sake of readability, I consolidate long, boring, intellectual property and plant industry terms. For example, "intellectual property" is abbreviated simply as IP and plant breeder's rights (PBRs), which in some countries may be referred to as plant variety protection (or PVPs), will be collectively referred to as PBRs. I will use the terms "cultivar" and "variety" interchangeably. Similarly, I will often imply PBRs to be included under the scope of *patents*, especially when I am speaking generally of "protection." When speaking of *trademarks*, please understand that trademark protection extends to both products and services. When I refer to "products" as it relates to trademarks, I am implying both products and services (unless otherwise indicated).

Any legal lingo I include is intended to improve communications between yourself and your IP professionals. For the record, I am not an attorney and I do not offer legal advice. The observations offered here are based on my experiences as an IP portfolio manager and as a self-employed business-person.

A word of caution: Trying to use legal jargon with your attorneys is potentially hazardous. Your attorneys will be thrilled to hear their native tongue spoken by you, will make assumptions about your level of understanding IP, and they will then proceed to rattle off a deluge of foreign terms at a dizzying pace. Be sure to promptly interrupt them, and make sure they clarify what they are saying and that they don't make assumptions about your situation – that is a golden rule of *POOF! Don't let IP professionals make assumptions about your goals, resources, and situation, or they may accidentally lead you down the wrong path.*

Exceptions Immobilize

In his upcoming health book, *Don't Take Advice from Fat Doctors! The Honesty Diet®*, Dr. Sean Hannon introduces what he refers to as the "Yeah, but..." syndrome. He explains that this universal syndrome can paralyze us from taking action because it is so easy to think of an exception to any suggestion or piece of advice we receive from others. We use this "Yeah, but, I know a guy who that didn't happen to..." reasoning as a thinly-veiled excuse for not taking action on something that on some level, somewhere deep down, if we are being honest with ourselves, we know we should do.

We can all cite exceptions to everything I propose in this book. These exceptions could be used to refute, remain immobilized, or otherwise fail to take corrective action. We all, for example, have an uncle who smoked, drank, ate three eggs and bacon everyday for breakfast, and is still alive and spry at ninety-five. *Your goal is to have an exceptional IP portfolio – not become the exception.* Exceptions are either lucky or expensive. I prefer to not push my luck and keep costs as low as possible.

Details Distract, Principles Empower

The stories presented in this book will not be fully presented. Frankly, who'd want to read every detail of any situation? Each client's circumstances and each plant asset are unique. I deliberately highlight portions of some stories for illustrative purposes and I deliberately ignore particular circumstances of other stories because, while their facts are likely quite interesting, their minutiae detracts from recognizing core principles regarding *how to think about and how to correctly apply* intellectual property.

This is a book that helps you understand core challenges with intellectual property in the horticulture industry so that if in the future someone challenges your legal rights, you can count on your solid business practices to substantiate (and therefore keep) your inventions and revenue.

Actual and Composite Stories

Some of the stories told in this book are true stories and include the real names of the participants based on publicly-available records. I may occasionally refer to facts stated in legal cases when telling a true story. However, the representation of these court cases is not intended to fully depict or interpret the facts surrounding each case.

Other stories are composite stories expressing real events that I have observed (sometimes repeatedly) in more than a decade of intellectual property experience. In these composite stories, the names have been changed and some circumstances have been merged with one another to facilitate storytelling. Collectively, they are examples of real problems faced by real companies today. Consider them the John and Jane Does of a communal intellectual property graveyard, if you will.

Disclaimer

Just as you should never begin an exercise program without first consulting your doctor, you are similarly and *strongly* encouraged to contact an IP professional, such as an attorney or patent agent, before applying anything you've read in this book to your specific situation.

CHAPTER ONE

Susan's Story: Trademark Limitations

"Hey, You Stole My Plant!"

Susan's passion for gardening even baffled the enthusiasts in her gardening club. She certainly wasn't a fair-weather horticulturist. Summer ice hailstorms may batter her broad-leafed plants, but under such circumstances, Susan resumed her work in her 16 x 24 foot greenhouse. Susan was most known for her miniature hostas.

Over the last two decades Susan had developed twenty-two new varieties, fourteen of which were still best-sellers. She sold the miniature hostas under her *federally* registered trademark PARAPET CROSSING®.[1] The name of which derived from her son's use of her hosta hedges to "protect" his toy soldiers during

1 This is a fictional story; therefore, PARAPET CROSSING is not an actual registered trademark.

pretend battles. Susan held this trademark for six years; in fact, she just finished paying her first trademark renewal. She was pleased with the trademark and many people in the industry recognized and liked the trademark, especially once they read about her son's inspiration when visiting her sales website.

Lying among her amassing pile of seedling trays was an invitation to speak at a national landscaping convention. Normally, Susan avoided such events because she hated to leave her plants in the care of anyone but herself. However, she was thinking about expanding her trademark into other landscaping products. She thought the convention might give her an opportunity to look for potential trademark licensing partners in order to create a full-fledged brand from her trademark.

A couple of months later Susan was walking among the vendor stands at the conference, when she recognized a display filled with miniature hostas. Upon closer inspection, she suddenly realized that several of the varieties for sale were her hostas! In talking to the salesperson, she learned that their distributor was someone in North Carolina, but Susan didn't recall any business relationships with anyone in North Carolina. Trying to remain anonymous to the salesperson, she extracted as much information as she could. She also noticed that the plants were sold under a different name – not PARAPET CROSSING®. Later, when she was discussing the theft over the phone with her husband, he assured her that they would contact their trademark attorney as soon as she returned home.

Susan's dilemma doesn't end happily. In fact, here is an example of a woman who had built a profitable, well-known business based on her trademark and her plant varieties. Unfortunately for Susan, she made a fatal flaw very early in her business development, and it wasn't until after meeting with her attorney that she learned her business was at risk of going *POOF!*

When Susan and her husband met with their attorney, she explained the problem and how she wanted to take immediate action. She wanted an *injunction*. "Isn't that what they call it?" she asked the attorney. "I want to stop them from selling my stolen plants!" Her attorney looked concerned. He asked her how many *patents* she had for her plants. "*Patents?*" she repeated, "I don't have *any* patents. I have this trademark." Now the attorney seemed even more concerned.

"Look Mark," Susan's husband interjected, "I know you didn't file Susan's trademark application. We filed it ourselves and it was only after we had trouble answering an office action from the Trademark Office that we brought it to your firm. Perhaps that is why you aren't too familiar with our situation, but surely there must be something we can do. Certainly, there must be recourse for blatant theft? Susan carefully uses the trademark, giving the public notice of her rights, just as you instructed us to do. Surely that will protect her business?"

The attorney then began to explain the limitations of trademark protection when it comes to plants. He explained that the protection extended to *the name,* PARAPET CROSSING®, or the trademark, but that Susan's legal rights didn't prevent others from using her hostas to propagate and sell the new plants. "Unfortunately," he said, "if these companies sold the plants under a different name, then there was no clear way to legally go after them." The only way Susan could prevent others from propagating and selling her plants was to obtain patent or plant breeder's rights protection. Susan and her husband were stunned into silence as they reflected on all the long hours Susan spent building her business and now recognizing just how vulnerable it was.

Branching Out or Going Out on a Limb?

It took some time and careful strategic planning with her IP professionals, but Susan recovered from this setback. She even wrote online about her experience and received many supportive comments from online readers. Several said they think that they may have bought her plants under a different name. "From now on," they said "they would only look for the PARAPET CROSSING® label, so they knew they had a 'Susan original.'" Susan was finally creating a customer-valued brand.

She also reviewed her plant inventory, but she was only able to file for a plant patent for one of her twenty-two plants because all the others had been sold and were in the public domain *for more than one year,* so they didn't qualify for patent protection.

Two weeks after an article recounting her experience and comeback was published in a notable nursery journal, Susan was approached with requests to license her trademark for several home and garden products. Remembering her IP consultant's mantra that "Business development is like painting a room, where most of the work is in the taping and setup," she made an appointment for a strategy session with her IP professional.

During her strategy session, Susan learned that her trademark was only applicable to just one set of goods – "Live Plants." She did not, in fact, have trademark protection for any other horticulture and garden-related products and services. That meant that a competitor could, in theory, obtain a trademark very similar to hers, but for a category of goods *other than live plants*. Once again, Susan was caught by surprise, but at least this time she was learning before there was an actual problem. She couldn't believe her business was *still* so vulnerable and never imagined that selling her plants would have so many complications.

Susan's IP consultant assured her that many companies are faced with the same sort of trademark limitations. Since you must demonstrate that you use the trademark in connection to a commercial activity, if you aren't actually selling a product or offering a service in a particular sector of goods or services, then, technically, you can't have federally registered trademark rights anyway. Also, the consultant offered, many companies, especially smaller ones like her own single-owner operation, had limited funds and each category of goods claimed cost more money to obtain and maintain protection. Together, Susan and the IP professional devised a strategy that expanded her trademark brand progressively, into categories where Susan had viable business opportunities through this new licensing contract. Susan didn't waste any time before filing another trademark application to expand into these new categories.

In many situations, trademark applications begin with a single category of goods or services. As the business grows, and once it makes sense, the scope of the trademark protection can be expanded. There are several strategies for handling trademark and brand extension, but for Susan, it made the most sense to obtain protection in additional categories of goods.

Managing Costs for Your Trademark Portfolio

Investing in a trademark requires more than just fronting the financial resources. If you aren't going to promote your trademark, then why spend the money to legally protect your right to use the trademark? This seems like an obvious question with an obvious answer until you consider the common human habit of buying items and then not using them. For example, how many of us have workout equipment in our garages, basements, and even under the bed? When was the last time you used this equipment? If you are like most people, the last time you used it was about two weeks after you purchased it.

Many trademarks experience the same neglected fate. If we accept this understanding and you have a limited trademark budget and limited marketing resources, just as most of us have limited attention spans for exercise, then an effective strategy is to buy one piece of equipment that provides a whole-body workout. You need something versatile to increase the likelihood of using it. The same is true for a trademark. The notion of what I call an *umbrella trademark*[2] is a trademark with broad claims to a larger category of products in the horticulture industry.

Susan used this idea of an umbrella trademark for all "Live Plants." The same trademark could then be

2 An "umbrella trademark" is not an officially recognized term, but is used here to better explain the concept.

applied to a number of plants demonstrating similar qualities or characteristics. The benefit of this approach is that while you can have protection for any number of plants, you only pay for one trademark. You also save yourself a lot of time and money in the process because, essentially, you only have one trademark to promote. This practice departs from the historical approach in the industry of assigning one trademark per plant. Considering trademark protection does not legally prevent others from propagating and selling your plant, under a different name, singular trademarks for singular products do not offer much advantage to anyone other than the intellectual property professionals helping you file for trademark applications and charging you fees accordingly.

Brand Dilution – The Pitfall of Brand Umbrellas

There are many elements to consider if you decide to expand your trademark into a brand, and then license that brand. They include:

- What kind of partners will you have?
- Do they fully support the values of your brand?
- Do they have systems in place to correctly use and monitor the use of your licensed IP?

- Will customers be confused by the brand extension?
- Are the products and services of the same quality that matches a consumer's expectation of your brand?

For many, brand licensing may seem like the pinnacle of success, but there is also a risk that it will destroy your hard-won reputation.

As Susan prepares to license her trademarked brand, she can learn from two of my favorite umbrella brands. They are Virgin, managed by English businessman, Sir Richard Branson, and Trump, managed by American businessman, Donald Trump. They are my favorite umbrella brands because they demonstrate the same approach yet yield very different results from one another. As business leaders, these two gentlemen successfully rule their respective billion-dollar empires. They both expanded their brands, but in my opinion, this strategy has worked well for one more than the other.

I am not sure why I like the Virgin brand so much, considering I am not really a customer, but I suspect it is because I *like* Branson, his business values, and the way he presents himself. He has created an umbrella brand valued at 24 billion dollars (2012)[3] for a menagerie of seemingly unrelated products – Virgin Records, Virgin Mobile, Virgin Atlantic (airline), and Virgin Galactic (space travel). The less known, and purportedly less successful, applications include:

3 Virgin. (2014). About Us. Retrieved on April 15, 2014 from https://www.virgin.com/about-us.

Virgin Brides, Virgin Cola, and Virgin Cosmetics.[4] According to Sir Branson himself, "We have created more billion-dollar companies in more sectors than any other company."[5]

So, from music to space travel, to anyone other than this child of the Sixties, perhaps this seems like a classic case of brand dilution. How can one company successfully protect the image of their brand across such different industries? The answer lies, in part, to the consistent application of a unifying thread, or in industry terms, the *brand promise*. A brand promise tells the customer that each and every time they use that brand product or service they will experience the stated promise. Promises can be difficult to live up to, but when you've managed to consistently keep your promise, you've managed to create an extremely valuable brand.

In an interview, Sir Branson summarized the brand promise of Virgin Group (the investment company name at the top of the umbrella brand food chain) as, "I believe the Virgin brand, for many people, now stands for a lifestyle choice that sums up our core brand values of *fun, innovation* and *value for money*, and it attracts people who appreciate and demand great customer service."[6] It seems that the customer-centric

---

4 Wachman, R. (January 7, 2012). What Does Richard Branson Really Own? *The Guardian*: The Observer. Retrieved on April 15, 2014 from http://www.theguardian.com/business/2012/jan/08/virgin-brands-richard-branson-owns.
5 Branson, R. (March 29, 2011). Richard Branson on Branding. *Entrepreneur*. Retrieved on April 15, 2014 from http://www.entrepreneur.com/article/219405.
6 SwitzwerDaily. (2013). Small Business: Brand Power with Richard Branson. Retrieved on April 15, 2014 from http://www.switzer.com.au/small-business/business-management/marketing/brand-power-with-richard-branson/.

combination of fun, innovation, and financial value can be applied to nearly every industry, so perhaps this is why the Virgin brand has been so successful. Sir Branson also attributed the success to a very strong and talented team.[7] Like Susan, as she was just beginning to launch her PARAPET CROSSING® brand, if you want to start licensing your brand, make sure your partners care about your brand as much as you do.

Without such a strong team or a universally-attractive brand promise, extending a brand into different product lines and services may only confuse customers because they don't expect to see your brand for that type of product. If your brand-product doesn't make sense, they will most likely ignore it. As an example, let us look at the Trump brand.

Donald Trump is an equally successful businessman. He doesn't strive for the "let's hang out and have fun" impression that Sir Branson exudes. Perhaps that is what isolates Trump from universal appeal. If I were doing a real estate development project, there is no one I would rather have on my team than Mr. Trump. My impression of his brand, reinforced by his TV show, *The Apprentice,* is that Mr. Trump is a man who can cut through the crap and get things done. But is that his brand promise? Other words that come to my mind, based on watching him on TV, are that his brand is synonymous with opulence. When I began writing about brand promises for this book, my impressions of

---

7 LeadersOnline. (October 4, 2009). Distinctly Entrepreneurial: An Interview with Sir Richard Branson, Founder of Virgin Group. Retrieved on April 15, 2014 from http://www.leadersmag.com/issues/2009.4_oct/eni/Branson.html.

his brand were congruent with the visual presentation of his famous apartment buildings, hotels, and casinos. But what about water?

Yes, that is right. Some of you may remember that Trump extended his brand into the bottled water industry. Why? Other than the fact that it was a rapidly growing market, I couldn't guess. Also, if he wanted to financially gain from this growth industry, why not just buy an existing company? Why use his own brand label?

After some research, I discovered that in addition to real estate and golf courses, the Trump brand was applied to the Miss Universe Pageant, home furniture, men's clothing, and the Trump Signature collection which included chocolates, tea, vodka, and, of course, water.[8] So how can we look at this brand extension and apply it to our own business or plant line? I think it is by learning what not to do. According to a Trump representative, "The Trump name speaks to indulgence and a luxurious lifestyle."[9] OK, so what does that have to do with a pageant for beautiful, talented women? Certainly Trump loves beautiful women, but a brand is a promise *to your customers*. So, as a customer, how will I experience indulgence and a luxurious lifestyle by watching the pageant? I won't. The brand promise fails to deliver.

Similarly, Susan's trademark, PARAPET CROSSING®, has some limitations. A parapet is a small, fortified wall

8 Klara, R. (May 2, 2011). Brand Trump: How the Developer-cum-TV-Star-cum-Presidential Candidate Became a Living Product. *Adweek*. Retrieved on April 15, 2014 from http://www.adweek.com/news/advertising-branding/brand-trump-131155?page=2.
9 Ibid.

intended to protect soldiers. This is fitting considering the shape of hosta hedges and her son's playful inspiration. But what value is that to me as a customer? Susan will be challenged to extend her brand, but because of the popularity of her plants and her respect in the industry, she has managed to attract parties that want to license her trademark. In fact, one mail-order company wants to use PARAPET CROSSING to market garden tools. Is this going to make sense to the customer? Will they identify a connection to the brand's promise? These are the challenges of transitioning a trademark into an umbrella brand. Unlike the Virgin brand, I think Susan will face challenges more similar to those of Mr. Trump, in that the brand can't be applied across the board and still be meaningful to consumers.

CHAPTER TWO

Jim's Story: Trademarks & Emotion

What's in a name? That which we call a rose
by any other name would smell as sweet.

—Juliet Capulet, *Romeo and Juliet*
play by William Shakespeare

Yeah, right! Anyone (such as a marketing or IP professional) who has ever tried to persuade a business owner to change their mind about a product, company, or brand name will not readily agree with Juliet's naiveté. It is true that a good name can really sell a project or product. However, that same name that you love can get you into real trouble, too.

Names are emotional, and therefore, names are powerful. People grow attached to names very quickly and are often reluctant to let go of them. Names help make something seem more real. They help us establish an emotional connection. When we know our waiter's name and use it, we tend to get better service. Early 20th century author, Dale Carnegie said that one powerful way to "win friends and influence people," is use their own name when speaking to them.

For example, on a funny sitcom television show from the 2000s called *Scrubs,* husband and wife characters Turk and Carla went into a panic about raising children after naming their unborn twins. Prior to naming them, they were excited about having twins. They romanticized about being parents and all the cute things the kids would do and all the wonderful experiences they would have as a family. Turk and Carla then came up with the names "Angie" and "George." After that, the panic set in. "What if George did this?" And, "What if Angie got sick?" These scary scenarios repeatedly woke them in the middle of the night. While naming something can give that something life, it can also imprison that life. *So...*

It Pays to Be Flexible

A name, especially at the early stages of a company or product development, isn't set in stone. Your customers do not yet have the same attachment to your name and this is evidenced by low or no sales. Your endearing name is simply a name, and it certainly isn't a brand yet. Perhaps wisdom does come from the mouths of babes and Juliet was correct – once we set aside our emotional attachments to a name, perhaps a different name would sound (or smell) just as sweet.

Ultimately, a name is intangible, and remembering such allows for flexibility while brainstorming. If you become too attached to a name early on, a name that your customers do not even know yet, you may set yourself up for serious risks. This is particularly true if your proposed name was to violate someone else's intellectual property or if the name were to create *customer confusion* as to whom the source of the product belongs.

Intellectual Property's Intimate Relationship to Marketing

Many clients are somewhat surprised when I tell them of the intimate relationship of a company's marketing efforts to their intellectual property. They often think these two departments are separate from one another and cannot impact one another. But the reality is that these two departments are like concentric circles, overlapping with one another and both departments must have a functional, open communication system between them. For example, when I worked with Western Union®, business managers often came to our department and excitedly told us about the name of a new product. Many times (and against policy) they had already created marketing materials with the proposed name. Sometimes they had even publicly promoted it to consumers.

Similarly, even large European horticulture companies that have been offering products since the mid-1800s can create a problematic disconnect between their marketing goals and their IP goals. In their excitement over a clever name, they often forget to consider the ramifications to the intellectual property of the name they chose.

If you intend to do business nationally or globally; if you intend to invest tens of thousands of dollars in advertising or an otherwise large percentage of your budget on promotional efforts; if you'd prefer not to find yourself facing a frivolous lawsuit; then you need to consider whether or not the name is "clear" for you to use in commerce. You must undertake a thorough effort to avoid weakening your brand and/or leading you towards a path of infringing on other people's legal rights.

And, on that note, you must understand that…

You Are the IP Police Force

Many people think they can file for a patent or a trademark and, once granted, their principal responsibility is to negotiate lucrative licensing royalties, followed by the purchase of a quaint Caribbean condominium. While these are exciting outcomes to possibly look forward to, it is critical to recognize that your intellectual property portfolio is not a passive asset. Rather, it is an active "muscle group" in your business that must be constantly exercised.

A country's Patent and Trademark Offices (including the U.S.) do not monitor potential infringement of your intellectual property assets. Instead, these offices rely on the "good faith" of the application filer (applicant). They trust that when you file an application, you've met all their rules and requirements. In most circumstances, and unless someone challenges your IP, these offices will not actually verify your statements or the statements of your competitors.

Once a grant is awarded, the Patent and Trademark Office's role in your application is essentially finished. No Patent and Trademark Office (as far as I've encountered) in any country employs enforcement police. *You* are your own IP police force. *You* are the one responsible for monitoring who uses your patented plants and who sells them. *You* are responsible for identifying who may be infringing upon your IP assets. *You* are the one responsible for correcting your licensees and horticulture magazine editors when they use your trademark incorrectly. No one else will do it for you.

Some government offices do try to help you. For example, you can notify the United States Customs Department that you have a federally registered trademark and they will do their best to confiscate any goods coming into the country illegally using the trademark. However, I have a feeling that while these hard-working and under-appreciated employees may be able to spot U.S. National Football League contraband, they may not have the same acute eye to recognize violations of your less recognizable plant trademark. Nonetheless, it is advantageous to have as many people assisting you as possible, so it doesn't hurt to notify them of your trademark rights.

Similarly, you notify the public of your rights through proper labeling. This means using the Register "R" symbol, ®, once you've received approval from a national Trademark Office to do so (not before); and using patent labeling language appropriately. Proper use of IP rights notification language will be explored later through the example of a company that experienced a terrible *POOF!*

For now, it is important to understand the legal and potential monetary value of IP notifications. If the public is notified of your legal claims to certain IP, then their future legal defense that they were "unaware" is less meaningful to judge or court during litigation. If it is proven that a third-party blatantly ignored your IP rights, well, then you may want to book an appointment with a Caribbean realtor.

Charbucks vs. Starbucks®

Nothing is so unproductive as the law.
It is expensive whether you win or lose.

—Gilbert Parker (1862-1932)
Novelist and Politician

"What's the big deal?" Jim Clark probably thought to himself. Jim is the CEO of Wolfe's Borough Coffee,[10] a small, New England-based bean roaster. Sure, Black Bear Charbucks Coffee Blend sounds an awful lot like the famous coffee company Starbucks. Both are coffee brands that rhyme with one another, and one can see an obvious reason why the name Charbucks attracted the attention of Starbucks' corporate legal team. But should Starbucks have legal claim over *any* words that might rhyme with their corporate name? What about "Four Bucks Coffee"? Would Starbucks have a problem with a competitor of this name, too?

10 Wolfe's Borough Coffee's is now doing business as Black Bear Micro Roastery – sourced from Nathan, D. (November 26, 2013). Wake Up and Smell the 'Charbucks' Starbucks Loses Trademark Appeal. Thomson Reuters: *The Knowledge Effect*. Retrieved on May 1, 2014 from http://blog.thomsonreuters.com/index.php/wake-smell-charbucks-starbucks-loses-trademark-appeal/.

Avoiding Customer Confusion

A primary goal of the Trademark Office is to sanction a commercial, competitive advantage for different forms of marketing and branding. They balance this support with the idea that use of a trademark should not "create confusion" in the mind of a customer.

In our Charbucks/Starbucks story, the question is: Will anyone mistakenly buy Charbucks Blend, believing that they are *actually* purchasing a Starbucks product? Wolfe's Borough Coffee didn't think so,[11] and in the end, the courts agreed. But that didn't mean success to Wolfe's Borough Coffee.

After the November 2013 court decision, interviewers at CNBC expected to hear elation from the CEO of Wolfe's Borough Coffee. What they received instead were comments from a man whose voice was so drained it sounded like he hadn't yet had his morning coffee! Jim's exact words were, "All my life, all I ever cared about was just getting this over with. I mean it's been a long hard fight and it's been an awful lot of stress."[12]

All his life? That wasn't much of an exaggeration. This court battle lasted twelve years, a good-sized chunk of Jim's professional career. Now, let me pose to

11 Black Bear Micro Roastery. (N.D.). What It's All About. Retrieved March 2, 2014 from http://www.blackbearcoffee.com/Starbucks/What's_it_all_about.htm.

12 Gusovsk, D. (November 18, 2013). Small New England Coffee Company Outlasts Starbucks. *CNBC*. Retrieved on March 3, 2014 from http://www.cnbc.com/id/101208003.

you a very important question: Do you presently have any brands, trademarks, or product names that you are so attached to that you'd be willing to turn over twelve years of your life to defend? Well, the answer to that question might depend on how valuable you believe that name to be and how profitable you think it could be.

Any Publicity Is Good Publicity?

Jim Clark and his company were judged to be in the right. Customers were *not* going to be confused by the name Charbucks, which was inspired by the bean's smoked, dark roasted characteristics. However, does being right equate to a worthwhile victory? Is being right a satisfactory exchange for the colossal distraction of a lawsuit?

Some people claim that any publicity is good publicity. *Really?* Does being the center of a public lawsuit necessarily increase your brand's visibility? If so, it will surely lead to an increase in sales, right? According to Jim, the answer for his company was a definitive, "No." When asked by interviewers whether the case increased his business, he conceded, "We get inundated with phone calls, interviews, and everything, *but it has not helped our business one little bit, not once in the twelve years.*"

Don't Be Too Clever

Let's all be honest for a moment (after all, no one is watching). Have you ever had an idea for a product name or trademark that was knowingly built upon the framework of an existing famous brand? Haven't you felt a little tempted to use this play on words, secretly hoping that prospective customers would make the connection between your modified trademark and the famous one? Of course, you have. I have, too. It's normal to think this way. It is particularly advantageous when trademarks rhyme, as was the case with Charbucks and Starbucks. Our nervous systems are programmed to recognize rhymes and patterns, and we like recognizing such.

When we brainstorm for new names, trademarks, and brands, it isn't uncommon to look outside ourselves for inspiration. What's worked well for others? What's catchy? Sometimes my clients will think of something that's a clever offshoot of a famous trademark, expression, or name. When they are in the middle of pitching their idea to me, I can usually tell that they already anticipate my objection. Their speech rapidly increases and their voice gets both louder and higher in pitch. They hope to fully express their argument as to why they have a good name before I interrupt and pose my, "How much is it worth to you?" question. Now, if my clients listened to and followed their gut instincts, then I would not have to stifle their excitement; but once we come up with a clever name, it can be hard to let it go. The truth is that somewhere in the recesses of our mind, we know that what we are proposing isn't the right course of action.

Similarly, isn't it probable that *someone* at Wolfe's Borough Coffee associated the famous Starbucks coffee trademark to the name Charbucks? Of course, they did. Again, while the courts eventually sided with Charbucks – that the similarity would not cause confusion to consumers – Jim clearly regrets the legal battle. It wasn't worth the fight.

Flying Under Radar? Or Crashing Into Low Bridges?

Would you agree that sometimes one can know just enough to get themselves into trouble? Many people think that they are a small enough business that they can fly under the radar. They argue, "I can do it until they tell me to stop." These people have been around enough attorneys to be familiar with terms like "Cease and Desist" and "Injunctive Relief," where essentially you continue with business as usual until you are caught. Then you receive a certified letter in the mail, a verbal slap on the wrist, and you are instructed to stop. Perhaps you've garnered enough sales and a large enough customer base to carry your business through this minor nuisance. You retract your marketing materials, change the name, and everyone continues on their merry way. Right? Not so fast.

Have you ever heard the expression: *It is better to ask for forgiveness than permission*? Many of us have and often operate on this premise. Of course, this doesn't make you a bad person. It is a natural thought process, but there are more consequences than people may think if these thoughts are acted upon. You may be inviting trouble to a party you really don't want to host.

Wolfe's Borough Coffee didn't have the opportunity to ask for forgiveness. They did not receive an order for "Injunctive Relief." Instead, Starbucks went straight ahead with a full lawsuit – one that drug out for twelve long, exhausting years. Not only did Charbucks not fly under radar, not only was the connection not missed by the Starbucks legal team, it was, in fact, Starbuck's *legal responsibility* to pursue Charbucks in order to enforce and maintain their trademark. We will discuss this concept more thoroughly in our next story. So, let me ask you one last time, "Is it worth it?"[13] [14]

Even if the courts eventually side with you, it doesn't mean that you've won. Going to court is exhaustive, expensive and distracting. Charbucks' legal victory was the exception, not the rule; and it was a victory with

13 Ibid.

14 There is certainly more to this story than is represented here. I've boiled down the sentiment of the CEO's statements to reiterate my point, which is to ask yourself "Is it worth it?" This is a particularly interesting trademark case. If you would like to read a statement about the origin of the name from Wolfe's Borough Coffee, you can access it at: http://www.blackbearcoffee.com/Starbucks/What's_it_all_about.htm.

no spoils. You may not think the innocent naming of a new product or brand could land you in court, but in the world of IP there are many opportunities to catch you unaware. Don't risk it. If you know your brand or product is riding on another's coat tails, hop off now!

CHAPTER THREE

Seth's Story: Trademark Dilution

Borrowing on the Identity of Others

I recently saw a newspaper advertisement for Seth's Plumbing. My eye was immediately drawn to the advertisement because their logo very closely resembled the famous "Superman" logo. Before I realized that the company was a plumbing company, I thought I was going to see something about Superman, a new movie perhaps?

Integrating well-known imagery into your marketing may be tempting under the hope that you are tapping into an existing, well-primed market association. In essence, you are benefitting from the marketing investments of another company; however, what are the ramifications of such a strategy?

The Power of Imagery and Their "Famous Brands"

Our minds are programmed to think rapidly by recognizing symbols and colors in order to create near instant mental associations. This is exactly why obtaining trademark protection and using trademarks to build strong, consistent branding is such an incredible tool. Just a few exposures to a distinguishing logo and our brains have the capacity to recognize the company forevermore. For example, we can recognize the blue sky and bright red and yellow colors of an old Western stagecoach as that of the Wells Fargo® bank logo even before we actually read the words "Wells Fargo." The leverage gained from immediately communicating without the use of words, converts thousands of marketing dollars into millions or even billions of dollars of market-value through brand recognition.

Small, local businesses like Seth's Plumbing seem to be notorious for taking a "few liberties" by associating their own small company, with a larger, recognizable brand, logo or slogan – by accident… of course. As for the close rendition of the Superman logo, the owner of the plumbing company knows darn well that anyone turning that newspaper page will likely notice the near-Superman logo and subsequently read the advertisement in order to determine what the ad is about. These newspaper readers may not have otherwise taken any notice of Seth's advertisement.

Business people like Seth know how difficult it is to get someone to look at an advertisement. Perhaps it is for this reason so many companies rely on pictures of puppies or babies. I mean, what does a chubby baby have to do with car tires? Yet the Michelin® baby was a signature marketing concept for years. The same can be said of the Taco Bell® chihuahua. Normally we do not want to associate dogs with our own meals, but Taco Bell made us all pay attention to their advertising and they even managed to make their small dog a small part of our Western culture for several years.

Establishing an association with an image that we recognize or like with a company's product, is considered a marketing homerun. Unless you use a generic, universally cute object, like a puppy, most of these mental associations require continuous public exposure, which costs a lot of money in advertising. Achievement of such a widely-recognized association among the general public has a legal classification at the U.S. Trademark Office referred to as a *famous mark*. The definition of a famous mark (trademark) may seem self-explanatory, but the ramifications of violating the legal rights of a *famous mark* are perhaps less understood and can be severe.

As for our "Superman" plumber, he certainly wanted to benefit from a famous brand. Seth also may contend that his ad was generally an innocent use of the Superman logo. Can he support this assertion? I said that the logo was *almost* like the Superman logo. If Seth felt his actions were entirely innocent, why wouldn't he use the *exact* logo? Perhaps he understood that what he was doing was not ethical, perhaps not even legal. His justification for his action most likely relied on the fact that he was just a small business advertising in a local newspaper. In other words, he was counting on the fact that he would never get caught by the Superman logo trademark owner. What could happen if Seth was incorrect in his gamble?

The Difference Between Bullying vs. Policing

As previously mentioned, governmental Trademark Offices do not employ a regulating mechanism to monitor who is using a trademark. A trademark owner does not get to simply file an application, use the trademark accordingly, and magically benefit from the legal rights associated with an approved trademark (called a *trademark registration*). It is the trademark owner's responsibility to closely monitor others who have permission to use the trademark, as well as those who do not have permission.

In other words, it is the trademark owner's responsibility to police the use of their trademark both internal and external to their organization. If the trademark owner fails to do so, or does not consistently "police" their trademark, surprisingly, the owners actually run the risk of losing some or all of their trademark rights. The owner of the Superman logo is a company called DC Comics®, and it is their responsibility to pursue Seth for trademark infringement.

Business news media is replete with stories of how one party violated the trademark rights of another party. This violation is referred to as *trademark infringement*. Particularly when it is a large company suing a small company, the public often criticizes the large company as "bullying" or unnecessarily attacking the smaller business like a predator. Is this bias a fair, educated judgment? Industry peers and consumers may not be aware of the legal expectations and responsibilities of a trademark owner. Therefore, they may mistakenly judge your actions as aggressive or over-reactive. Such a judgment may be inferred from the Charbucks vs. Starkbucks® story presented earlier. Understanding the expectation of the court system can help us all realize that these trademark lawsuits, generally speaking, are not frivolous. They are a way to prevent trademark dilution, and are, in fact, an expectation of the courts.

Continuing with the bullying or predator-prey metaphor in order to demonstrate the importance of protecting your trademark rights, let us imagine the trademark owner as a mother cheetah. The cheetah exerted herself, using her cunning and strength, and investing her time to kill a large animal to feed her offspring. These activities can be likened to the efforts of filing for a trademark, obtaining legal rights, and building an associated brand.

Based on the size of the animal the cheetah killed, she has the resources to feed her three small cubs for weeks, but in order to do so, she knows that she must cover the animal carcass with dirt and hope that no other animals take notice while she retrieves her cubs from their hiding place. The cheetah securing her food, while she attends to other matters, is akin to establishing a set of "trademark use" guidelines that explain how to properly use the registered trademark. This is, in essence, establishing a territory and public claim to the trademark. Cheetahs are lone hunters, which presents a real challenge for a mother cheetah that must hunt and protect her family at the same time. Many trademark owners face a similar constraint in the amount of resources that they can devote towards protecting their trademark rights.

As the cheetah and her family return to the buried food, they notice that hyenas have arrived (i.e. trademark infringers). The mother cheetah is distracted by her need to protect her cubs first (i.e. distracted by the running of the business), and so she decides to merely sit and watch the hyenas, rather than immediately chase them off. At first, the hyenas take a couple of small bites because they want to see what the reaction of the cheetah will be. The cheetah doesn't attack, instead, she keeps watching. The hyenas are emboldened, so they take larger bites and begin howling (i.e. infringing on a trademark through their marketing efforts). The howling attracts several other hyenas who rush in to take advantage of an undefended, free meal. If the cheetah were to continue to observe instead of act and defend her kill, she would watch her earned source of food get eaten away, piece by piece, until nothing but un-palatable bones remained.

Slowly Eroding Your Rights (...And, Your Future Profits!)

In this same manner, a court that would normally support your assertion to the rights of a registered trademark, will consider the *entire scenario* of the trademark infringement case. Questions they will ask include: When were you first aware that someone was violating your trademark rights? What did you do about it? Did you decide that the infringement was just a small incident (just a couple of hyenas taking few small bites)? Did you decide that you were either too busy, or didn't otherwise have the resources to assert your claim over your trademark? Perhaps you didn't want to incur legal costs for an attorney to write a *Cease and Desist* letter? Or perhaps you felt too uncomfortable at the idea of confronting the infringer? Maybe you didn't want your actions to be misunderstood by others in the industry, which may result in labeling your company as an over-reactive bully. So you decided to turn a blind eye to the problem.

Your actions, no matter the rationale, will be judged by the court system evaluating your trademark infringement case. In other words, it isn't just the actions of the infringing party that matter. If your actions do not indicate to the court that you value your trademark and you fail to defend your trademark against numerous attacks, then this behavior supports the legal designation of *trademark dilution*. In the courts' mind, why should *it* support your legal claims to a trademark if you can't demonstrate that *you* even support your trademark rights?

Trademark dilution can also happen internally within an organization. Trademark protection only extends to the precise representation of the trademark as presented to the Trademark Office in the registered application.

If you happen to trademark a phrase, protection extends only to that *exact* phrase. If you utilize multiple variations of this phrase in your marketing; you may be undermining the value of that trademark and thereby diluting your own rights. The same can be said of images. Trademark owners often encounter this challenge when they obtain trademark registration for a particular logo and then, after using it for years, they (often for reasons little more than boredom) alter the logo. When it is time to renew their trademark rights, they are surprised when the Trademark Office rejects their example of their revised logo because it does not exactly match what was approved. Dilution, whether it happens within or outside of the organization, is a common pitfall for many trademark owners and it can lead to a trademark's value going *POOF!*

Observe Your Competition – Then Be Different

Founder of WalMart®, Sam Walton, was notorious for his habit of patrolling the aisles of his competition. During his observation tours, he took detailed notes on what worked well for his competitors and what didn't. What kind of inventory were his competitors selling? How did they organize their store? How did they price their products? "How can I do it better?" he asked. This last question allowed Sam Walton and his company to become one of the largest companies in the world in a remarkably short period of time.

Primary Objective: Avoiding Customer Confusion

Analyzing your competition is a beneficial business practice and most of us do it, though sometimes not as deliberately as Sam Walton did. You may start this analysis by asking which competitor appears most similar to you? How do your tag lines, web addresses, and color schemes differ from that of your competition? If you sell the same products, is there a chance that your website is so similar to your competition that a customer may actually buy from you by mistake? This scenario, when two companies have such remarkably similar marketing materials or customer interfaces that the consumer can't distinguish the difference between the different companies, is an example of what the Trademark Office considers *customer confusion*. Avoiding consumer confusion is the primary objective of the Trademark Office and they take this duty very seriously.

When you file a trademark application, the Trademark Office will conduct a thorough search in multiple databases to try and determine if your trademark is unique from existing trademarks. If the Trademark Examiner determines that one or more existing trademarks are too similar to the trademark listed in your application, then they will reject your application because they do not want to create a situation that can lead to consumer confusion.

Therefore, when you are conducting your "Sam Walton" research, the goal is to be thorough in your observations, draw conclusions on how you can make improvements on what your competition is doing – but then make a 180 degree turn by creating marketing materials, customer interfaces, and trademarks that are unique to your organization.

The concern of creating customer confusion extends beyond the purview of trademark examiners and the concept is also strongly supported by the court system. If you are facing a trademark infringement lawsuit, and your competition can demonstrate that you deliberately copied elements of their protected trademark in an effort to intentionally deceive the public, well then, you are in a lot of legal trouble, and the financial damages assessed by the courts will be significantly higher than if your situation were merely a case of "innocent infringement."

CHAPTER FOUR

Tool: Avoiding Trademark Rejection

So How Do You Come Up with a Good Name?

One of the most valuable suggestions I can offer you is this: If you are currently using or want to create a trademark name or brand that sounds or looks similar to an existing one – don't. Instead, be more creative.

We've looked at several examples of what *not* to do and you should by now have a good understanding of the kinds of situations the Trademark Office wants to avoid. But how can you come up with a good trademark that will likely be approved? Fortunately, the Trademark Office offers several guidelines. With the help of your IP professional, combine the following guidelines with what is commonly called a *trademark name consultation* in order to seamlessly integrate the trademark with your business strategy. Either conduct a trademark search or have your IP professional conduct one for you and you are off to a running start.

What are you really asking for when filing for a trademark application? In exchange for you applying for a federally registered trademark, the Trademark Office will grant you *Exclusive Legal Rights* to the use of the trademark *if they consider the trademark unique*. Without the uniqueness criteria, the Trademark Office would end up establishing legal barriers preventing everyone from freely using common words in the English language without paying a royalty!

Meet the Trademark Office Examiner… The Decider

When thinking about a new trademark, it is helpful to understand the expectations of the Trademark Office. Your application will be reviewed by a randomly-designated attorney at the United States Patent and Trademark Office (USPTO). I believe Trademark Offices in other countries follow similar procedures, though, of course, their laws and requirements will vary.

Generally speaking, this attorney is the one person who will solely determine whether or not your trademark can receive federal legal rights. Of course, there are caveats to that last statement, but by focusing on the objectives of that one attorney, called an *Examiner*, you can improve your application's conduciveness for approval; thereby lowering costs and speeding up the application process.

In essence, *you only have one individual to persuade* the value of your trademark, which means that receiving federal trademark protection, is in many ways *subjective*. The Examiner does use court case decisions and existing rules and procedures to guide their decisions, but if your trademark follows guiding principles, you have a much greater chance to receive approval.

Trademarks Cannot Be Generic

A critical rule of thumb is that *you can't receive trademark protection for a generic or common word.* This is because trademark protection is designed to allow you to receive a monetary gain through *Exclusive Rights,* which may manifest in the form of a license agreement, increased consumer sales attributable to brand awareness, and other market-place advantages. The government can't allow people to file lawsuits or charge extra royalties for words that are part of our common language. The definition of *generic* encompasses a broad number of variables. Unless you want to receive a prompt rejection from the Trademark Office and you have plenty of extra money to burn up on unnecessary IP professional fees, then you will want to file a trademark application that does *not* break any one of these general rules:

1. The trademark cannot be a cultivar name. The Trademark Office considers a cultivar name to be the common name, therefore, it cannot be trademarked.

2. The trademark should not be a descriptive word or phrase. A red rose cannot receive trademark protection for RED ROSE®[15] nor can the trademark include a description of a preferential quality, such as BEST JUNIPER®. Descriptive names may be helpful for communicating in marketing materials, but the Trademark Office does not want to grant *exclusive rights* to the use of descriptive words.

3. The trademark cannot be descriptive in another language. A work-around trademark submitted to try and trick the Trademark Office cannot be a descriptive term in another language. In other words, ROSA ROJA®, which is Spanish for red rose, will be rejected by the Trademark Office. America is home to many languages. The Trademark Office recognizes these languages as "common" words that can't be trademarked. This even holds true for languages less prevalent in the United States, such as Russian, Japanese, and even Latin.

4. The trademark should not include the name of the type of plant or category of service. Sometimes these applications are approved but they are not what are considered "strong trademarks," which means if a competitor really wants to challenge

15 The "registered trademarks listed in this section are fictional, unless they are otherwise indicated by the trademark owner's name.

you in court, you may find out that the trademark rights you paid for at the Trademark Office might not be strong enough to provide you the legal advantage you were expecting. For example, a trademark for MANDARIN SUNSET® for a mandarin tree may actually be approved by the Trademark Office, but the office will not give you "Exclusive Rights" to the word "mandarin." Without exclusive rights, you will have a weaker trademark. Another possible example is to trademark the name of your nursery, say, FIREBIRD NURSERY®. In that case, you would not have exclusive rights to the word "nursery."

5. The trademark should not include the geographic location where you found the plant. This introduces all kinds of arguments from the Trademark Office, which will drive up your application costs in a hurry. If you file an application for NAPA VALLEY GRAPES® for example, you will receive a rejection from the Office, which may be very difficult to overcome.

6. The trademark should not be a person's name. Adding your name to the trademark, for example, JOE® or MILLER® can create a number of rejections from the trademark examiner. In some cases, this can be overcome, depending on the situation, but generally, unless you have a very distinctive name, it is best to avoid trying to trademark a personal name.

What Names Do Examiners Like?

Examiners are looking for something *unique* and *distinctive*. In the horticulture industry, this can be quite a challenge because, as you know, your colleagues are excellent at assigning creative, colorful names for plants. If these names are too prevalent or the name for a plant is too similar to the trademark you want to use for your plant or line of plants, then it is likely that the examiner will not consider the trademark listed in your application as unique. So how can you make your trademark more unique?

1. Employ fanciful words. That is the phrase the trademark industry uses. The challenging question is: Can you create a surprising combination of words to express the qualities you are trying to capture with your trademark? A wonderful example is SuperCal®, for a line of hybrid plants of petunia and calibrachoa, owned by Sakata Seed America, Inc. Other fanciful words are when you create words that didn't exist before, such as WIKIPEDIA®. Many technology companies favor this approach and while the Trademark Office likes these kinds of trademarks, because the word is made up, it may require quite a lot of marketing dollars to educate the public about the name. There

are also other considerations from a marketing standpoint that may make such trademarks difficult to market.

2. Unexpected combinations. APPLE COMPUTERS® holds a number of variations of trademarks because the combination of "apple" with "computer," at the time, was unexpected. Another unexpected combination could be a trademark for BARBWIRE PETALS® for a line of delicate, flowering plants.

3. Create a unique spelling. Changing the spelling or presentation of letters can be enough to make a trademark unique as long as it isn't merely changing an "s" to a "z" to avoid looking like a trademark someone else already owns. One popular trademark that uses this strategy is Toys "R" Us®, where the trademark has quote marks around the R.

4. File a trademark for a stylized word design or picture with words (a.k.a. Design Mark). The Trademark Office loves these because the picture creates a distinct representation of the trademark.

5. Be the first to use the trademark. In the trademark world, the "Exclusive Rights" prize is awarded to the one who uses the trademark first in the market-place.

Are You The First?

To test out potential trademarks, conduct an online search. This is a quick way to see if there is existing competition. Use your trademark with related industry words such as "plant" and "nursery." What results do you get? Are there any sites that come up with the name and/or offer similar services? How similar? How many uses? If you select this trademark and spend money to advertise it, you wouldn't want your prospective customers to inadvertently go to one of your competitor's websites or storefront thinking they were responding to *your* print advertisement, would you?

CHAPTER FIVE

Margaret's Story: The Price of Information

It isn't easy or cheap to find out how to obtain intellectual property protection for plants, especially international protection. Internet information is not always reliable and rarely comprehensive. Information found on the Internet often addresses only the most common elements associated with obtaining legal protection and many organizations have unique circumstances affecting how and when they should file applications. Government sites can be very helpful, but they also have limitations. For example, they only list the government fees and typically do not offer any fee ranges on what costs you can expect from local IP professionals to prepare the applications.

Establishing Oneself in a Global Market

There is a prestige to being able to that say your products are available in key markets around the world. In some situations you may want to work with partner growers in Ecuador, Mexico, China, and Bolivia and then sell your product in Europe and the United States. Or, perhaps you think there might be a great market in India and you want to grow the product in Turkey.

It is common that growers are not located in the same countries as your key markets. The strategies vary based on what partners you have, how much money you have, how viable the plant or seed itself is, and how realistic it is that you can fulfill the demand in a particular international market. We can see how these questions play out through Margaret's story. Margaret's experience will probably sound somewhat ludicrous and completely preventable, but believe it or not, this sort of thing happens all the time. I've witnessed it on at least a half-dozen occasions.

Runaway Costs and Information Acquisition

Margaret is the Director of Intellectual Property for a small seed company in the United States. Both she and the company have been positioning themselves to either grow the company based on their own products, or position the company to be bought by one of the larger seed distribution companies. Noticing the current upswing in company mergers, the owners of the company are favoring a buy-out and their new line of hybrids seem like the ticket to a well-deserved retirement.

Margaret is charged with keeping as many market and IP opportunities available for as long as possible, until there is a cash-infusion from investors to support a more robust global strategy.

Margaret didn't get to enjoy the quiet morning she expected. With both her bosses at a seed conference in Rotterdam, The Netherlands, she planned on a productive morning to catch up on paperwork. Instead, she was bombarded with emails and frantic voicemails from Roy, one of her bosses and the co-owner of the company. He became convinced by a conference speaker of the necessity to file plant breeder's rights (PBR) applications in several countries in Africa. He needed to know what the procedures were and how much it would cost. Furthermore, he needed this information in time for his meeting with potential investment partners the following Tuesday.

Margaret immediately telephoned her company's local attorney and asked that he contact representatives in Africa to obtain the urgently requested information. The attorney, who was well-versed in helping Margaret's company with their legal contracts, but less experienced with the ins and outs of international PBR law, contacted a large law firm who had offices in several of the countries Margaret mentioned. Even though the weekend was about to begin by the time the U.S. attorney spoke with the Moroccan attorney, the law firm assured the U.S. attorney that they could deliver the requested information on time; but that, naturally, there would be "rush fees" assessed because of the last minute request. Indeed, true to their word, cost figures and filing details were provided to Margaret in time for her to prepare her boss for the Tuesday meeting.

The Information Gap

There exists a large gap in expectations between clients and IP professionals. Many clients think conversations with IP professionals, such as attorneys, should be at no cost. After all, they were just having a "quick" conversation, there wasn't any actual work being done, right? Wrong. Conversation is the form of work that IP professionals engage in. IP professionals are in the information business. Information is the *currency* with which they have to trade goods and services. Clients often call up their attorney or send them an e-mail with "just a quick question" but then they resent receiving a bill for maybe $175 or more. What I have come to realize is "I just have a quick question" is a code word for "I don't want to pay you for information." But this is an unrealistic expectation to have.

IP professionals earn based on time invested and there are only so many hours in the day. In other words, their capacity to earn is limited by available time and by what rates clients are willing to pay. Also, to IP professionals, there is rarely such a thing as a "quick question." What clients think are simple questions, often are not; and may, in fact, require a lengthy, sophisticated answer, which often requires time-consuming research from the IP professional. Imagine an IP professional addressing dozens of "quick questions" from clients each day at no compensation. They'd never earn anything and would quickly go out of business! It may seem unfair but there is a high price to pay for experience. Keep in mind, though, that in the long-run, utilizing someone else's hard-earned experience will actually reduce your overall costs and lessen your chances of error.

A week later, Roy returned from the conference having already dismissed the idea of expanding into Africa. The investment partner was not interested in growing the seed in Africa and, therefore, would not bear the costs. However, at Roy's urgent instruction, Margaret had already organized all the emails that came in from each of the African attorneys and plant breeder's rights agents who responded to her inquiry. The bills for the research began to come in and, with mortification, Margret calculated the total at $6,587.00 USD. *What!* She thought to herself, *how is this possible? We didn't even file any applications, yet!*

Now Margaret was faced with the horrible position of explaining to her boss that the company had just spent almost seven thousand dollars for some emails about procedures in countries that, in some cases, didn't even have plant-specific protections! Given her bosses' abrupt and temperamental disposition, she knew that she couldn't rationalize the expenditure on his rushed demands and impulsive questions.

Further, this wasn't the first time she'd received an unexpectedly high legal bill from her U.S. attorney. Based on the kinds of questions he would occasionally ask her, she suspected that some of his billable time was accountable to the fact that the attorney wasn't familiar enough with their industry and, in fact, he was charging hourly to conduct basic research for Margaret. In other words, the company had to pay an inexperienced attorney to "get up to speed" on these issues in order to answer her questions.

When she asked questions about how to file for PBRs in the different countries, she, perhaps naively, never expected to be charged for her questions from the agents in Africa and yet that was the largest portion of the bill.

In replaying the sequence of events, Margaret would admit that her company did place a high demand on receiving the information quickly and her boss wanted answers to some questions that were specific to their situation. When she extrapolated the questions to six different countries, and that each of those countries have their own legal systems and distinct procedures, Margaret could now see why they received an invoice but she never imagined it would (or should) cost this much – and there was no way of accurately predicting this amount.

Balancing Curiosity with Strategic Planning

It sounds like a lack of common sense from Margaret's boss, right? What was he thinking giving Margaret instructions like that? He should have known how expensive that research could be. Well, I would agree with you that the error was on the part of Margaret's boss; however, this story is not uncommon. Obtaining information from attorneys (especially international ones) can come with a hefty price tag. Sometimes we think we are asking very simple questions, in which case, we don't expect to be charged, or at most, be charged very little. However, as I've mentioned, there are very few straight-forward answers in intellectual property, and so no IP professional should send you a cookie-cutter, stock answer.

The answers to questions, even seemingly simple questions like "When do I need to file this application based on my first sales date?" varies widely, depending on the country, type of plant, and the circumstances of an "offer for sale."

These questions must be carefully considered by an IP professional before they provide an answer – and, this takes time and money. Should it take $7,000? Not if you have the right IP professionals on your team.

Any IP professional with international experience knows that costs can quickly get out of control once you traverse your own country's borders. These professionals have either already developed relationships with trustworthy individuals who take more care in not incurring excessive legal expense, or they have experience in establishing cost stipulations before they even ask international legal counsel a question.

How to Control Fact Finding Costs

To avoid Margaret and her company's *POOF!* experience, or at least mitigate it, there are several things you can do.

Tip One: There's no such thing as a "quick" intellectual property question

Your IP professional will ask you more questions, spend more time thinking about your questions, conduct some research, and prepare a thoughtful response. Prepare yourself for the fact that they will likely bill you for all this time.

Tip Two: Estimate ahead of time how much it may cost to obtain the information

Instruct your IP professional that *before* anyone spends more than a specified amount, they must first receive approval from you or you will not be responsible for the charges. Make sure all this is communicated *in writing* and in the event that any legal professional challenges you on this, be ready to enforce the agreement.

Tip Three: Request data (i.e. "order work") in progressive stages

Sure, it would be really nice to understand how much IP costs in each of the countries you have a fleeting interest in and what the procedures are, but do it in stages so as to control and monitor costs. It can be very difficult to find accurate, comprehensive information online. Certainly you need to understand at least the basic requirements for certain countries and this information is not always readily available online, but it can also take considerable time for the IP professional to pull this kind of information together for you. Again, this is because the information depends largely on your specific circumstances.

I will give you a practical example. Let's say you would like to obtain plant breeder's rights (PBR) protection in Australia. Since Australia is an island, they are particularly sensitive to the types of plant material imported into their country. You may have a variety that Australia considers invasive and, therefore, they may not allow the plant to be imported. Each country has their own particular regulations and, furthermore, these regulations vary depending on the crop. You may have no problems importing some plants into Australia, but then suddenly run into a brick wall with another plant.

If you need detailed information in order to develop your strategy, then again, take the process in stages and try to keep everyone on a budget.

Tip Four: Have more than one IP professional on your team

As wonderful as it would be to only count on one person to help you manage your IP, you will definitely need more than one professional to help you in this highly specialized field. My wonderful patent attorney friends may not like me revealing the fact that no singular IP professional can meet all your needs – even if they have a science degree in horticulture, but it is true.

For example, nearly every patent attorney in the U.S. will tell you they can file a plant patent. I know this because nearly every attorney biography I've read states this. Technically speaking, they are accurate in claiming their permitted scope of practice, but nearly every patent attorney grossly underestimates the underlying complexity of plant patents and many, if not most, lack the experience. Having only one IP professional on your team will *necessarily* mean that you are paying him or her to get educated on your unique circumstances instead of using someone that already has the expertise and can get straight to the work at hand. This is a waste of your valuable resources!

The applications themselves aren't necessarily overly complicated, but the circumstances surrounding plant-related patent applications are nearly *always* complicated.

There simply is not one single attorney or law firm who has all the expertise you will need to successfully manage your IP portfolio. There will be times when you need specific scientific expertise, or litigation advice, or company merger experts, international contract attorneys, or PBR experts. This is why I work with a referral network of intellectual property professionals and scientists from several sources; because I've never found a single organization with all the experts my clients need for each of their unique situations.

Tip Five: Organize the data you paid for in one location

All of your IP portfolio data must be accessible from one location or from one resource. Make sure anyone who may need this information to make a decision (i.e. spend money by ordering work and asking "quick questions") knows of and has access to this data.

A friend of mine owns an information technology software company. He recently received a telephone call from a large corporation who said they would like to purchase his software program. "I am sorry, I can't sell it to you," he said.

"What?" the stunned corporate buyer asked.

My friend went on to explain, "I will not sell it to you because I already sold this system to your company five years ago. I will; however, come out to your office, retrain your employees, and update the system to meet your current needs. Will that work for you?"

You can see why this person of high-integrity is my friend. He didn't take advantage of a company who lost track of the software they had already paid tens of thousands of dollars for. Not every other provider is so conscientious. Ironically, the software he sells is a knowledge-management system!

Keeping track of the information you pay for is a hassle – at best. I realize that there are very few people in the horticulture industry who like to spend time on the computer, but failing to assemble the information into one location and then sharing it with the appropriate people (especially as new employees take on new positions) is critical and will save you a lot of money.

One simple system is to copy the text for the inquiry emails you receive into one large text document or spreadsheet. The information doesn't even need to be overly organized. The next time you want to find out if you can file a PBR in Ghana, then conduct a search for the word "Ghana" in your Word document.

CHAPTER SIX

Yoshi's Story: Data Access & Open Communication

I cannot emphasize enough how perilous patent and plant breeder's rights (PBR) filing dates are. They are one of the main reasons why a patent application is either denied at a Patent Office or rescinded through the court system after granting. In most cases, the problem lies with the fact that the right hand does not understand why the left hand is doing what it is doing. A lack of communication, including an up-to-date plant portfolio tracking system, leads to lost opportunities, lost revenue, and unnecessary expense.

Yoshi manages the IP portfolio for a small, horticulture research company. As with Margaret in our previous story, Yoshi, too, is in charge of obtaining PBR protection for several lines of hybrids. The success of these varieties solidifies the success of the company. Yoshi's company has spent many long, stressful years growing, and is now on the verge of international expansion.

While Yoshi enjoys going out to the test plots and taking characteristic measurements alongside the breeders, he ultimately considers himself an office person. As such, he prides himself in his masterpiece - his cultivar database. Using the spreadsheet, he tracks the different research names of each variety assigned at each stage of development, and he keeps up-to-date records on research articles relevant to their propagation methods. The spreadsheet also includes charts of each genealogical lineage and the marketing department provides Yoshi the dates of first sale for each variety so he can calculate what the filing dates will be if they elect to file in particular countries.

Operating under a completely different set of objectives and responsibilities is Janet, the company owner. She is a natural-born sales person who also manages all the company's contracts because the contracts reveal sensitive financial information that Janet prefers to keep private from her employees. Despite Janet's access restrictions as it relates to the company contracts, Yoshi and Janet enjoy an open, cordial working relationship.

Upon Janet's request, Yoshi was preparing the paperwork for two PBR applications for a new, hardy, market-worthy hybrid. One application was for the U.S., the other for Canada.[16] What Janet and Yoshi had

16 Some countries refer to plant breeder's rights as PBRs, while other countries use the term plant variety protection or PVP. While the names differ, the types of protection they afford are essentially the same.

yet to realize was that her "simple" request to submit two new PBR applications, would open a Pandora's Box of IP complications as it relates to plant protection.

A few days later, Yoshi entered Janet's office to inform Janet that because the patentable plant was a hybrid, they are required to submit *both* the new hybrid's seed samples *and* also those of the hybrid's inbred, parental lines. While this requirement was familiar to Yoshi, Janet was somewhat taken aback. The request triggered a concern regarding an obscure recollection somewhere in the corner of Janet's mind that the history of the parental lines may pose some complication for the PBR applications.

Once Yoshi left her office, Janet reviewed the private contract files for the parental lines. Yes, there it was, just as she recalled. Her company had purchased the parental lines from her brother's company, which was located in the Niagara Lake region of Ontario, Canada. The first sale was completed, as indicated by the contract, just *ten months prior* to today's date. Whew! Janet was relieved to confirm that the parental lines were sold to her *within the required one-year sales date stipulation* under U.S. plant variety protection laws. Like most humans who confirm their hopes through some "quick research" and then promptly stop looking into the matter further, Janet confidently closed the file and stopped thinking about this hybrid's history. That was Janet's mistake.

Yoshi finished preparing both the U.S. and the Canadian applications without incident and paid all of the fees on behalf of the company to the PBR agent they hired. In due course, the applications were both granted and the company did, in fact, generate significant revenues from this hardy hybrid. The future was promising for Janet's company and the hybrid's sales in the U.S. and Canada soared. As a result of its aggressive market performance, this popular hybrid then drew the attention of a cut-throat Canadian competitor.

The competitor held a close relationship with someone at Janet's brother's Niagara nursery. Under a series of circumstances that can only be attributed to Murphy's Law, the competitor somehow gained knowledge that Janet and her brother held a contract containing an unusual and poorly-worded contractual mechanism of automated "offers for sale" between their mutual research projects. The hardy hybrid for which Yoshi filed a Canadian PBR fell under this contractual arrangement. The competitor, sensing an opportunity to go head-to-head with Janet's company's new hybrid sales, immediately challenged the validity of the PBR arguing in court that the hybrid had *technically* been offered for sale *prior to the Canadian PBR application filing date,* which, unlike many other countries including the U.S., is not permitted under Canadian law.

The Aftermath – Who's to Blame?

The facts presented in this story are multi-faceted for a reason. In most situations, a plant's history is, in fact, complicated. Yoshi's story illustrates that while *any* U.S. patent attorney or patent agent can *technically* file for plant IP protection, the validity of said protection extends well beyond a list of seemingly straight-forward and required plant characteristic descriptions. Only those with horticulture experience and related business savvy can successfully guide their clients through the frequently murky waters of the meandering river of intellectual property for plants.

The first threat that placed the marketable hybrid in peril was the relationship between Yoshi and Janet. On most points, their relationship was great; however, Janet was reluctant to share some critical contractual information with Yoshi or any one of her employees, even though she relied on Yoshi to guide her and her company in making prudent IP decisions. In fact, one could easily argue that Yoshi and Janet were misaligned. Here, the right hand didn't understand *why* the left hand was doing what it was doing. It would never occur to Yoshi to discuss Canada's unique requirement that a PBR application must be filed before any offer for sale is made because he was completely unaware of the unusual contractual arrangement between Janet and her brother's Canadian company.

Similarly, Janet was completely unaware of Canada's unique requirement because, as the business owner with different responsibilities, she didn't have the time to know specific PBR requirements for each of the countries she wanted to file applications in. After all, keeping track of such information was what she paid Yoshi for! It was this communication breakdown which eventually led to the complete loss of the hybrid's IP rights in Canada. An entire IP marketing opportunity had just gone *POOF!* all because Janet was reluctant to trust Yoshi with data that, as it turns out, was vital to his successful job performance. All the time and the thousands of dollars invested in preparing and filing the Canadian application was wasted unnecessarily – and, of course, much more was lost from the now lower royalty rate because there wasn't PBR protection and exclusivity.

When Yoshi was first made aware of the Canadian PBR challenge, he felt the full responsibility of failure. It was only as the legal case dragged on that he eventually acknowledged to himself that he did the very best he could given the information he had access to. Unfortunately, as an employee who took great pride in his work, this didn't make him feel much better.

The second lesson of this story is that the decision makers of an organization not only include the business owners or managers themselves, they also include the expert staff (both internal and external to the organization) preparing and managing the applications, for example, in this story, Yoshi and the Canadian PBR agent. It takes an open and comprehensive exchange of communication to successfully manage a plant IP portfolio. Yoshi knew the *facts*. He even knew about the unique Canadian requirement, but only Janet fully knew the complete *circumstances*. Yoshi couldn't possibly understand the circumstances *behind* the facts and *how* the circumstances related each fact to one another. Since Yoshi wasn't privy to the contextual information that made the facts *tell the whole story* behind the new hybrid, the company ended up unnecessarily wasting thousands of dollars in resources and revenues.

Business owners must realize that once you elect to file IP internationally, the complexity of your portfolio increases exponentially and requires far more monitoring. *If you only tell your IP professionals half the story, you risk receiving only half the protection* – or, in Yoshi's case, even less!

Canada offers an intriguing caveat to Yoshi's story because at the time of this book's publication, Canada holds a unique requirement that a PBR application does not benefit from a one-year grace period between the first date of sale (or offer for sale) and the filing date. Furthermore, Canada is not unique in adopting its own interpretations and variations of the internationally-recognized treaty for plant breeder's rights protection. So, it would be unfortunate and foolish for any of us to make sweeping assumptions regarding what different countries allowed as it pertains to plant IP protection. Our best safeguard for executing profitable global strategies is to maintain open, accurate access to all data that may affect PBR requirements such as filing dates.

Be First to File… Or Be the Last in Line

While discussing Yoshi's dilemma, this may be a good time to introduce another risk of international business and the cost of not anticipating potential outcomes. Again, this is something that I have witnessed on multiple occasions and have warned my clients about.

Many countries have "first-to-file" laws where patent or trademark rights are awarded to the first person to file with the respective Trademark or Patent Office. This presents a real challenge because even if you don't plan on taking your invention or trademark global – *someone else may do it anyway!*

Therefore, your product, brand, or company name is not always legally available for use in another country. In some countries where the IP rights of others are weakly enforced, it is common to, in essence, pirate someone else's trademark or company name. For example, it is *extremely* common to visit such countries for industry trade shows, pass out your brochures and business cards and then later be notified by a law firm in that country that your business or product name was subsequently registered as a trademark and that there is a website domain name address using your company name!

This egregious behavior seems rather unbelievable, doesn't it? Certainly there are mechanisms in place to protect foreign business people, right? Of course there are, but they aren't convenient or cost-effective. First, you would need to hire an attorney local to the pirating company's country to file protests with the appropriate government offices. Then, you will also want to hire your own attorney in your own country of primary business to act as a liaison because these are not straight-forward transactions. I use the word transaction, because that is all this really is - an exchange of money between parties - and you might not even have a viable business opportunity in that particular country! This is your recourse and it is cumbersome, indeed.

In addition to the cost, such legal protests can take years to run through the system. In situations such as these, the courts will in most cases side in your favor because they recognize the mal-intent of the other party, but that doesn't mean you receive a refund on all of your accrued legal fees and you will certainly not get your time back. This is a reality of global business.

What is important to understand here is that, for most countries, whether you are interested in trademarks or patents, if you are not the first to file, you may find yourself last in line. This is something to be aware of and to be considered *before* it happens to you. So, plan ahead on how you might respond.

CHAPTER SEVEN

Walter's Story: Inventors & Discoverers

Like It or Not… Intellectual Property Is Here to Stay

Whether you support the notion of applying patents and trademarks towards plants or not, the use of intellectual property in the horticulture and agriculture industry is here to stay. No matter your opinions on the ethical arguments behind fair-use, saved-seed, patenting life-forms, or using intellectual property to extract commercial value out of research and development; we aren't going to change the tide of intellectual property, which more and more breeders are now utilizing.

Intellectual property is a tool. Like any tool, if it is the right one, it can make your life infinitely easier. Or, if it is the wrong tool, IP can cut your business like a knife. Some readers have their doubts about whether or not to use IP to protect their plant inventions, and to those readers I say, given the risks accompanying poor IP planning, you are correct to doubt! So, it is best for all of us at least learn how the tool works.

Fair Use vs. Intellectual Property

Plant breeders who take the time to produce plants, but don't obtain IP legal protection for those plants, are more like artisans contributing to our "common use" knowledge. The use of their inventions is fair game to anyone who would like to use it. Some plant breeders think that all plants should be in the public domain so they are happy to contribute to fair use.

As such, it is not uncommon for people in the industry to share plant material. In fact, some might argue this practice is the backbone of the industry. But where lays the difference between sharing and theft? Where's the line, between what is allowed under IP laws and what is ethical? What kind of behavior do you want to engage in, and how will this affect your brand image in the industry?

Track the Breeding History

Whenever you apply for a patent or PBR protection, your application requires you to state the *breeding history* all the way back to the first publicly available plant. You are not required to state the breeding history beyond this point. If a plant that is legally protected is included somewhere in the breeding history and that same plant is owned by someone else, but they did not grant permission to use the plant, then your newly invented plant was created *illegally* and is not patentable. The issue becomes even more complicated because of our global commerce environment.

Let's say a plant breeder develops a superior quinoa grain in Peru that has twice the protein content of any other known in the species. This breeder has the resources to commercially protect their plant with a plant breeder's rights application. This plant breeder now feels secure about offering the quinoa grain for sale. Meanwhile, health nuts in the U.S. find out about the higher protein content of this particular variety and immediately favor this new variety over any others in the market-place. To take advantage of this new demand, a second company, with more financial resources and distribution systems than our Peruvian plant breeder, decides to capitalize on this market opportunity.

Their legal team discovers that while the plant is protected by plant breeder's rights, its protection is limited to just one PBR in Peru. To circumvent the PBR's protection, the second company simply takes the Peruvian quinoa across the border to Columbia, and pays farmers to grow out the plant locally. Production is successful and because this company has a lot of resources, they apply for plant protection in a number of countries, excluding Peru. Now, obviously, this is pure theft but that does not necessarily make it illegal. Intellectual property laws are country specific and protection is technically limited within the boundaries of the country offering legal rights. Would you consider this unethical? These are the unfortunate realities of a global market-place.

It is illogical to try to file for patent protection in every country. Only companies the scale of MERCK® pharmaceutical and Microsoft® and perhaps a few large seed companies have the resources to file for patent and trademark protection in multiple countries. Even with a thoughtful strategy, if your budget has limits, then it is best to select the core markets where you know you have the partners in place to extract reasonable financial value and try to ignore the other counties.

Incidentally, in some cases, patents and PBRs could be challenged based on the question of *inventorship*, that is, if you were willing to invest the resources to challenge the pirating company. Naturally, that involves hiring local legal counsel and the costs could explode, but some plants are worth the fight. Many are not.

Is It Even Patentable? Wild vs. Cultivated

Most people in the industry seem to be familiar with the restriction that a plant cannot merely be found in the wild and then be patented. If a plant is discovered, it must be discovered in a *cultivated area*. You may also think you understand the meaning of the term "cultivated area," but do you?

Plant breeders often become so excited about a newly discovered plant that they rush into the patent process before they meet all the criteria of a plant patent. It can be terribly frustrating and expensive if a plant patent application you filed for is outright denied. It is always imperative that a company do everything in its power to prevent this from happening. Like most of the scenarios in this book, the following story starts out under seemingly innocent pretenses.

A 2012 court case[17] confirmed that if all the criteria for obtaining a patent are not met, the patent becomes invalid. Sometimes an application will be approved by the Patent Office inadvertently, based on the information the inventor and/or patent owner provided (in good faith). Sometimes the actual facts listed on the patent are mistakenly listed. Your granted patent can be subject to invalidity at any time during the term of the grant.

17 United States Court of Appeals for the Federal Circuit Decision Nos. 2011–1459, 2011–1460, 6 August 2012 Retrieved on August 1, 2012 from http://e-foia.uspto.gov/Foia/ReterivePdf?system=FCA&flNm=11-1459_1.

Patenting Plants Qualifications

There are two ways to qualify as an inventor for a plant patent at the U.S. Patent Office.

1. A plant's genome is altered through human intervention, which results in new and distinct characteristics. Once this is achieved, an inventor must then prove these characteristics are stable and that propagation is reasonably repeatable.

2. A new plant is discovered in a cultivated area by a potential inventor; identifies a new characteristic expression of the plant and then reproduces those changes reliably through asexual propagation.

A person cannot claim to be an inventor if they merely discover something in nature – in that case, they are a "discoverer;" but the discovery is *not* patentable.[18]

"I Can Get a Patent for This, Right?"

So what *exactly* does the U.S. Patent Office mean by a cultivated area? Unfortunately, the Patent Office fails to

18 The US Patent and Trademark Office provides full guidelines (under 35 U.S.C. § 161), which may be accessed at: http://www.uspto.gov/web/offices/pac/plant/.

define "cultivated," however, the true story of Walter and the August 2012 court ruling does provide some insight.

Walter Beineke filed two plant patent applications for two oak trees. Walter was pleased with the superior quality of his oak trees because of their excellent timber quality and strong central stem tendency. The patent application examiner at the U.S. Patent Office reviewed the statements Walter provided in his application regarding how he came to discover the oaks and the examiner decided that the area of discovery was, in fact, not a "cultivated" area as Walter stated and as the law required. Walter was very surprised[19] by the examiner's assertion because after all, he found the trees in someone's front lawn – surely that is a cultivated area!

Strongly believing that his inventions should be awarded patent rights by the Patent Office, Walter took his patent application to an appeals board within the Patent Office. This federal court reviewed Walter's case and, in the end, agreed with the patent examiner in denying Walter patent protection. Unfortunately, while Walter still had his wonderful trees that he could sell, he lost the opportunity for higher licensing rates that patent protection can offer. He also incurred lots of IP professional expenses with little to show for his efforts.

19 Walter's surprise is conjecture on the part of the author for story-telling purposes. The full details of this case are not represented here.

Where Did Walter Go Wrong?

Walter discovered the trees in a yard. During the hearing it was established that the trees likely started growing in the 1880s or earlier, whereas the house (and presumably the cultivated yard) was not constructed until the 1930s. The trees and the history of how they came to be predate the establishment of the cultivated area. This fact, or rather the lack of facts and the unknown, disqualified Walter's invention in the eyes of the Patent Office.

How can you make sure your breeding programs and policies do not create the same kind of mistakes that Walter's did? How can you make sure you do not spend thousands of dollars pursuing patent protection when the Patent Office will not be able to identify an actual invention? If Walter had some good analysis questions on determining cultivated vs. wild locations, he could have saved himself thousands of dollars and tremendous frustration.

Questions for Discovered Plants

Plant patent applications rely on complete descriptions of the plant, and the area and circumstances of discovery should not be an exception. Taking a few extra moments to jot down descriptive notes of the area when a plant is discovered could mean the difference between later obtaining a patent or not.

Integrating the following types of questions into your plant discovery data assessments can help ensure that you clearly demonstrate to the examiner that the plant was discovered in a cultivated area.

1. Exactly where was the plant discovered?
2. When was the plant discovered?
3. Do you have pictures of the area at the time of discovery?
4. What other plants are in the cultivated area?
5. Which plants are likely to be the parents?
6. What do the possible sibling seedlings look like?
7. How are seedlings different from the discovered plant?
8. How is the area of discovery manipulated through human intervention? Fertilization? Irrigation? Selection and thinning?
9. Why was the cultivated area established?
10. How is the cultivated area maintained?

CHAPTER EIGHT

Raymond's Story: Patenting Isn't for Sissies

There is a big difference between plant breeding for commercial purposes and plant breeding for patenting plants. Patenting is a separate branch of your business, which means there are unique business procedures to follow. There are business behaviors to engage in and, of course, it means there is a need for systems and paperwork. If you don't think you want to devote the resources to turn patented products into a viable branch of your business, it may be best to not use this intellectual property tool. As you'll see from Raymond's story, patenting isn't something to just dabble in.

Raymond never thought he would be a patent owner. He never saw much value in it and now he wished he would have listened to his gut instinct. They say aging isn't for sissies, but neither is patenting. As the largest grower in Central Valley, Raymond knew something of intellectual property, mostly from the licensee side. He paid royalties to seed companies for the rights to grow and then sell several hybridized vegetables. He was successful in selling to the commercial market but his younger brother, Jaime, was encouraging him to expand his operations into breeding the varieties.

Jaime argued that Raymond was leaving too much money on the table and that the real money was in breeding. Jaime contended that with a breeding business they didn't need to maintain as much land to make higher profit margins. The only reason Raymond listened to his brother instead of his gut feelings was because, for the last two years, Raymond was being forced by the State to leave a portion of his land fallow so that the neighboring towns could use his water. Coming from a family of farmers, he knew occasional droughts afflicted business and you just had to ride it through. But Raymond worried about keeping his year-round staff employed while maintaining his vast plots of unused land. If he had to sell the land, and the drought and water restrictions continued, would there even be any buyers? Diversifying his business in a way that didn't increase his demand for land seemed like an intelligent solution.

The trouble was that Raymond didn't know much about breeding new varieties. As a grower he was restricted from deliberately trying to alter the seed he used. Jaime assured Raymond that he would lead the breeding program and that he had the perfect candidate for their first cultivar.

Creating Leverage with Patents

Jaime, a recent graduate of a prestigious agricultural college, like all young men, was full of fresh ideas. He knew that the best way to be competitive in this new market economy was to "do more with less." He'd been attending several business seminars and the speakers talked about leverage and efficiency to the extent that someone could dramatically reduce their workweek to just a few hours. Jaime grew up working long hours on the farm but now he could see that there was a better way to do things. He could even revolutionize the agriculture industry with these tools of leverage. Convincing Raymond to expand into plant breeding was just the first step. Jaime knew that in less than ten years they could move the operations to large greenhouses, employ just a few key people, file patents, collect royalties, and let someone else worry about growing their varieties. No more managing a large seasonal staff and no more crippling water restrictions. They could start selling some of the family land to the home developers who kept approaching Raymond.

Although Jaime didn't share all his plans with his brother (that would come in time), he promised Raymond that he would develop the first cultivars. The truth was Jaime had already begun long before he approached Raymond. A few summers prior, Jaime was an intern at a testing facility that was in partnership with his school. They had a lot of seed under development and countless others still waiting.

One senior graduate student leading Jaime's work claimed to have discovered a large green pepper with high water content. He asked Jaime to test the pepper without providing any further guidance. Jaime spent the rest of the summer measuring and testing, meanwhile, the senior graduate student never gave Jaime or the pepper another thought. Jaime returned for his internship the next summer to discover that no one had done anything with his research. He was frustrated, but just as in the previous year, he was left to his own devices, so even though no one was there to really teach him, through trial and error, Jaime learned a lot about breeding and the pepper was proving promising.

A couple of years passed and Jaime graduated. He was now working in sales at a nursery. He was tired of the office day-to-day grind and was complaining to his friend who still worked at the testing facility. "You should come in and volunteer on Saturdays. We could certainly use all the help we can get," said his friend. Jaime agreed, and he was happy to be back in the field. Jaime then noticed that the senior people still hadn't done anything with his pepper. Obviously the university did not see the pepper's real value.

Jaime was irritated that his hard work continued to go unnoticed. Considering the university had so many varieties to worry about, Jaime rationalized that he was the only one who did any work to improve the pepper, and he thought they would never notice a few missing seeds. He decided to continue his research on the green pepper, but this time, on his brother's farm.

Such was the testimony Jaime provided to Raymond's patent litigation attorney during a practice run of Jaime's deposition. Raymond couldn't believe what he was hearing. He knew that the deposition was a fact finding mission, but it was only after the university presented the confidentiality agreement to Raymond that Jaime disclosed the true origin of the seeds. The university claimed that Jaime stole the seeds and now Raymond realized with a sinking feeling that they were right. Raymond had been paying expert attorneys to defend his company and their pepper patent for the last three years. He'd long since exhausted his savings, and just as Jaime had hoped, Raymond did start selling off portions of the family farm, but that was just to pay the legal bills that were mounting. He could never imagine that his patent would end up costing him so damn much! Even as Raymond watched his brother in disbelief, Jaime continued to claim that he was the true inventor of the pepper.

Unbeknownst to Jaime, the university had finished testing the same pepper, but they carried on the preliminary work that Jaime had done during his internship in another facility off campus. Furthermore, they filed patents in the US, Japan, Australia and plant breeder's rights applications in Mexico and Argentina.

The university was having large commercial success with the same pepper as Raymond's company.

Perhaps no one would have even noticed that they were selling the same pepper, but Raymond's company used a marketing name that, as it turned out, included some of the *variety's denomination name*. Jaime, of course, recommended the name and it was this name that drew the attention of the university.

Years later, Raymond thought back upon all of the legal haggling. It was so distracting that he ended up leaving most of the farm operating responsibilities solely on his wife and her sister. The genetic testing of the seeds provided by Jaime proved that the seeds covered both by Raymond's *utility patent* and the ones file by the university had the same key genetic markers. Yet, Jaime insisted that he was the true inventor. After all, he was the one who bred and tested the final cultivar. Furthermore, considering the fact that Jaime's name didn't appear on the university's patents, this was, in Jaime's mind, enough to prove that the university's patent were invalid. Jaime wanted to counter-sue the university. For a while, Raymond's attorney reasoned that such could be a recourse to follow, but Raymond didn't want to be in the business of asserting his patent. He wanted to get back to the more tangible and straightforward business of growing plants.

The mere mention that Jaime could be a rightful inventor of the pepper encouraged the university's employment office to dig deep in their files. They found an *internship employee agreement* that automatically assigned Jaime's legal claim to any acts of invention to the university as a mere matter of employment. The university's Office of Technology Transfer, which handles the intellectual property protection and commercialization of the university's inventions, were relieved to hear that their older contracts for internships, such as the one Jaime signed, had an *Employee Assigns* clause. For Raymond, he realized that unless he was committed to developing a business around plant breeding and patenting, merely dabbling with patents could cost a fortune.

Patenting Pitfalls: Inventorship & Ownership

This story presents a number of common pitfalls in the patenting process. As you can see, it is relatively easy to make several mistakes before you even know you are making them. Perhaps the greatest lesson Raymond learned is that filing for a patent application and receiving a granted patent is, in some cases, just the beginning.

Filing *utility patents* for sexually-reproduced plants and their seeds, like this green pepper, generally costs quite a bit more in legal and administrative fees than a *plant patent* because the *prosecution* (the process of getting the application granted) is more complicated. Raymond factored these expenses into his cost-benefit analysis before he agreed to start a breeding and patent program. Unfortunately, his patent attorney didn't recognize that there may be uncertainties of *inventorship* and *ownership* of the patent. Without recognizing such problems, the patent attorney didn't feel it necessary to discuss some of the risks and patent enforcement responsibilities a patent owner may face. Considering that this was their first patent, neither Raymond nor Jaime could have anticipated the problems that lay ahead of them.

Obviously one can lay blame on Jaime, but in practicality, his attitude is not that uncommon and Jaime certainly felt justified in his actions. He didn't feel as though he stole anything. If you are running a plant breeding company, you can expect that at least some of your employees feel similarly. Your employees are likely providing expertise that you don't have or don't have time to pursue because you are running the business. But from the employee's perspective, they are the creator and the genius, if you will, behind the product. As they create more exceptional products for the company, it is human nature to feel justified for more recognition and higher compensation.

Relating to Jaime's situation, is it possible that your organization may have had one or two who behaved just as Jaime did? The main way to mitigate these problems is through good systems. Do you have systems in place to keep track of your plant material? Do you even know of all the different testing projects your employees are engaged in? Do some employees work on *their own* side projects *in your facilities*, using *your resources*? If the employees are using your facilities to develop their own projects, is it *clearly articulated in a written document* who owns the inventions created? Do you have employment agreements in place with all employees that include assignment of invention clauses.

Inventing and Owning: Not the Same Things

The question of inventorship and ownership, particularly in the plant breeding world, is complicated. In fact, this is the area that I see as the weakest in plant patenting systems and it leads to the invalidity of patents. The problem, as I see it, is that most people are only looking at a situation from *their* vantage point. A talented, employed inventor who solves previously unsolvable problems in order to create a plant of high market value, has a hard time thinking of their inventions as "all in a day's work." Like any creative person, they have a certain level of *perceived ownership* in the project. Perception of ownership leads to feelings of entitlement, whether they are an employee or not. Organizations that value their employees' contributions recognize this need, and they often develop an *inventor reward program*. The reward program can take many forms. I worked for a company that offered a fixed sum for every patent application filed. This sum was divided among the number of inventors per patent. In some cases the plant may not be patented, but the inventors may still share in a royalty program.

We all want our employees to perform above and beyond the call of duty. Driving innovation in your organization for the long term means that you also have a financial compensation program in place to encourage inventors to be happy to continue to apply their talents towards the growth of the company. In Jaime's story, he felt very isolated. No one directed his work, much less paid attention. Even though he solved critical problems for the university, Jaime didn't receive any recognition. This led Jaime to feel justified in taking a few seeds that he thought no one would notice were missing. While this was technically a criminal act, it doesn't necessarily make Jaime a criminal. It means that he was seeking confirmation of his significant contribution by any means necessary. If we think about it, we've probably either been in a similar situation or at least rooted for the "bad guy" in a good bank heist movie. These kinds of employee-employer dynamics can be avoided with fair recognition programs. So that is from the employee-inventor perspective, now what about the company, patent-owner perspective?

Many employee-inventors, because they are focused on their own contributions, fail to understand and recognize the significant business risks of the employer. Inventors are often artists, and as artists, they often have no idea what it takes to pull together an entire company, keep products moving so they generate cash flow, and meet employee payrolls and other debts. I've worked with both artists and inventors for many years and no matter how talented they are, I find they grossly underestimate the gap between producing a great product and successfully selling a great product. As an employee-inventor, it is important to recognize that, in most cases, your organization is not making nearly as much money at the end of the day (net profit) as it may appear or be assumed. There are numerous operational costs that employees rarely consider that greatly affect the profitability of a company.

It is also important for the employee-inventor to understand that in the absence of a written agreement that stipulates otherwise, and in nearly all cases, anything you create and invent while an employee *belongs to the company*. Statements of such an arrangement may have been included in the employee contract paperwork you signed when you joined the company or took the internship. You may not even remember, realize, or have carefully read what you signed. If it isn't written in the contract, and if it is taken to court, most courts favor an *implied* relationship between employer-employee, which considers the facts that you used the employer's resources to develop the invention, therefore, the *employer owns the rights*. Absolutely, as the employee-inventor, you provided the expertise, which is a vital resource, but generally speaking, unless you've obtained an agreement from your employer that says otherwise, the courts consider your genius as something that should be brought with you to work and used each day as an expectation of your employment.

Naming Inventors

The above discussion explores the human dynamics and contractual situation as it relates to invention ownership, but who is the actual inventor? The answer to this question depends on what country you are filing an application in. For example, you can file a European Community Variety Rights application and not list the inventor. By contrast, in the U.S. the inventor *must* be listed. Inventors are the individual(s) who managed and directed *most* of the activity during the development of the invention. If we look at Jaime's story again, we know of at least two inventors. The first is the senior graduate student who discovered the pepper. The second inventor is Jaime. Jaime is correct that he should be listed as an inventor, but that doesn't mean he automatically receives financial compensation. Additional inventors should be added depending on who was working on the project at the other university facility. All the invention work does not need to be done at the same place and at the same time. This is where we find a common pitfall for patenting plants.

Plant breeding projects can last for years. I was astounded when I was once told by a breeder that he'd been working on a tree variety for twelve years. If this invention is part of a larger breeding program, it is likely that contributors will come and go over the lifespan of the project. If, however, some have already left by the time the patent application is filed, that doesn't mean that they should not be identified as inventors. Anyone who contributes significantly to overcoming the development and propagation problems of the plant may be considered an inventor. A distinction is made by the U.S. Patent Office for anyone working under the direction of a lead project manager. Let's say that the senior graduate student directed Jaime in Jaime's work. The graduate student was involved in the different characteristic selection decisions and other critical decisions. Under these circumstances, one could argue that Jaime was not an inventor; he was merely following the guidelines set forth by the senior graduate student (i.e. the true inventor).

If a plant breeder inventor is not listed in the patent, then this is a problem. The Patent Office doesn't have the resources to investigate the completeness of those listed on the application. However, should the need arise; the Patent Office has investigative processes available to anyone objecting to the inventorship.

Inventor Assignments: Plan for Divorce While You're Still in Love

Whether you are working within an organization or a partnership, deciding whether there is going to be a delineation between inventor and owner for a patent is a conversation that is best had *in the beginning*. Taking the time to clearly identify roles and expectations between inventors can often avert future misunderstandings and frustrations. Discussing individual expectations helps inventors clarify the goals of obtaining intellectual property protection. In order to design a solid contract defining the working relationship, it pays to think ahead about what problems you would like to avoid, identify areas where you have mutual and different goals, and how you are going to establish a system for fair, transparent reporting and compensation.

In addition to the employee contracts (if you are in that type of situation) is a common legal instrument, called the *Inventor Assignment*. These are often filed with a patent or plant breeder's rights application when all the inventors are not necessarily owners. Individual partnerships may also use an Assignment in situations where there is more than one inventor, but only one inventor takes the financial and legal risks to file the patent application.

An Inventor Assignment is a legal document with important legal consequences and an attorney should be consulted before signing. However, if you are an employee, it may have been part of your employee contract that you agree to automatically assign your rights to any inventions that may be filed for a patent or PBR. In which case, the attorney you consult would tell you that you are under an obligation to sign the assignment. There are circumstances when employees, or ex-employees, refuse to sign the assignment. However, this does not preclude the organization from being granted a patent.

It is important to understand that an assignment does not mean that the person signing the document is no longer considered an inventor. An assignment merely identifies ownership of the patent. It is common in partnerships that there may be multiple inventors, but only one or two inventors are the owners. This type of arrangement encourages collaboration between breeders and then involves the inventors only to the extent they want to be involved when it comes to the business and IP side of the patented invention.

Based on these discussions, you can see why Raymond was correct to be concerned about starting a plant breeding program that includes patenting. There are so many details that unless you have the experience, or devote the resources both to IP professionals and to systems to manage all these details, patenting can result in a net loss instead of a net gain. That doesn't mean, however, that many organizations, who invest in these systems and education, do not learn how to make this leveraged tool work well for them.

CHAPTER NINE

Tool: IP Risk Assessment

Identifying Your Risk Tolerance

You're never going to get where you want to go if you do not first know exactly where you are. Please take the time to answer the following questions to gain valuable insight into your IP portfolio. Your answers will help you determine when or if it is intelligent to commit financial resources toward protecting a plant asset. Just by completing this assessment you will, believe it or not, have done more than most of those in the horticulture industry utilizing IP ever have.

1. Do I have the ownership rights to the plant material I developed? For example:
   a. Do I have a lineage with patent-protected material that may be owned by someone else?
   b. Did I inherit the plant material from someone else who may be the only inventor?

   c. Am I an employee or partner who is obligated under a contract to assign my claims of the invention to someone else?
   d. Did I discover this plant in the wild?
2. How difficult is this plant to propagate?
3. Is the plant material distinct, uniform, and stable?
4. How difficult is this plant to ship to distributors?
5. How many plants or seeds do I expect to sell?
6. How many seasons do I think the market will purchase my plant material?
7. How much money do I think I can "net" from the sale of each plant material unit?
8. Does this plant material meet a consumer's need or desire?
9. Is this invention representative of the core of my business? Or, is it a one-and-only endeavor?
10. Can I put a team in place to help me propagate, distribute, and market my invention?
11. How much money am I willing to invest in obtaining IP protection?
12. Should I bring in a partner to help pay for the IP costs?
13. Is it possible that my company name or product name may violate someone else's rights (i.e. trademark infringement)?
14. Am I willing to create paper and computer systems to track my plant breeding program and materials?
15. Am I willing to ask my partners to sign legal contracts *before* they work with me on my projects?
16. How different is my marketing and branding from my competition or famous trademarks?

17. Am I willing to spend the money to legally defend my rights against others?
18. If so, how much money am I willing to spend defending my rights?
19. Do I need both national and international protection? If so, when?
20. Is it financially intelligent for me to spend money to obtain legal protection and enforce my rights in *each* country I intend to grow and sell in?

Please discuss the results of your assessment with someone. Now is the time to talk out-loud, *before* you spend a lot of money on IP protection fees. If you do nothing more than answer these questions, at least you've told your brain what your IP parameters are. Your brain, which is constantly working to protect you and help you avert loss, will send urgent messages to communicate with your gut instincts. In fact, the most important principle in avoiding the pitfalls of experiencing a *POOF!* is simply to really listen to your gut. *If it doesn't feel right, don't do it – or at least wait.* If you are excited about your new plant, take a moment to talk with a vested, but less emotionally-involved third-party. This is often a significant other, close family member, friend, trusted colleague, or any other risk-adverse person.

Listening to your gut can keep you out of the court system, out of bankruptcy, and thus free up time and resources to be outside in the fresh air (where most of us prefer to be), developing new plants, or otherwise enjoying our lives.

CHAPTER TEN

Erik's Story: Consequences of Disorganization

Labeling products, price sheets, and marketing materials with "patent pending," a patent number, or other labels such as PPAF (plant patent applied for) and PVR (plant variety rights obtained) is a notice of legal rights and claim to the ownership of plant material. These rights, however, have expiration dates, at which time the public has full rights to freely propagate your plant materials. It is important to keep track of these expiration dates and to stop labeling the plant material as "patented" once the patent expires.

Accusations of Deceptive Marketing

Erik had been a territory sales manager for a European vegetable company for twenty-seven years. He loved his job and was excellent at it. The company had great products and Erik was a master at negotiating prime placement in supermarkets across the United States. Prior to taking this position, he worked at the company headquarters managing their intellectual property.

While sales were still strong, some in the company began to worry about losing market-share to the smaller companies who were coming out with new and more unique cultivars. Still, Erik's company had a strong brand that many consumers preferred and the company held over forty U.S. patents and twenty trademarks. Erik himself was in charge of coordinating the filing of most of the company's patents.

Anna, who represented a plant breeder's cooperative, was one of those newer competitors Erik's company had concern about. After losing a critical negotiation with a large grower, Anna asked the grower why they had not selected her organization since the co-op clearly had superior cultivars to Erik's company. The grower, who had always liked Anna, disclosed that the grower made more money with the products from Erik's company because their patents yielded higher licensing royalties.

Anna was surprised. *Hadn't those patents expired a long time ago?* she thought.

Years ago, Anna had worked an internship with Erik's company so she was familiar with the European company and their products. She had it on good authority that Erik's company hadn't developed anything new in a while and they had scaled back their patent program. Knowing this, she asked the grower if he would be willing to tell her which patents covered Erik's company's products. The grower was willing to share this information with her because, after all, the more he promoted the patented cultivars, the more money he made. So he gave her a copy of the latest catalogue and sales sheet.

Many of the board members in Anna's co-op disliked patents so when she reported that they lost a large account because of Erik's company's patents and that Anna thought at least some of those patents should be expired by now, they encouraged her to investigate further. Anna easily discovered that ten of those patents had expired. Yet Erik's company still labeled the products as "patented." Anna took this information to her board of directors and they saw this as an opportunity to finally nudge Erik's company out of their long-standing, dominant market position. They filed a lawsuit under the claim of "False Marking" of products.

As part of the final court decision, the judge agreed that gross neglect had occurred on the part of Erik's company. Considering Erik's role with the company as both an IP manager and sales manager, there was even a strong suspicion that the company intentionally neglected to update their product literature and materials to remove the patent labeling. This would represent a deliberate effort to maintain an unfair market advantage and reap the financial rewards of higher license royalties.

Track the Details – Stay Out of Court

False marking of products was a section of the patent law that was, until recently, rarely enforced. The problem with false marking, for example, by failing to remove your patented label after it has expired, as we saw with Erik's company, is that you are misleading the public into thinking you have legal rights that you no longer possess. In the eyes of the Patent Office, you are impeding innovation because people assume your patent isn't in the public domain, and so they do not include the now-expired patent in their research and development.

The Patent Office offers a deal to a patent holder – they'll concede exclusive rights of an invention to a patent owner for a period of time *if* the patent owner reveals in complete detail how the invention works. The principle is that complete disclosure will spur innovation once the invention is in the public domain.

This way others can benefit in the future from the inventive progress the patent holder achieved.

In the past, the deterrent for false marking was insufficient, because in many cases, the courts would only financially penalize a company *per incident,* not for each item that was mis-marked. Many companies failed to keep up with this administrative task because they assumed the marketing benefits outweighed any legal repercussions. Now the circumstances have changed, perhaps because of the aggressive competition in the electronics and software industries. Now patent law allows for a fine of $500 *per falsely labeled product*. If thousands or tens of thousands of products are in distribution at any given time, the penalty could be significant.

Good Faith

As we have discussed, the entire IP system is based on the premise of *good faith*. In historical terms, you may call it a "gentleman's agreement," where your word, in this case, all the statements you make to the Patent and Trademark Office, is considered true until someone (usually a competitor) challenges your statement and there is an investigation (either through the patent and trademark system or through legal courts). Certainly there are people who don't adhere to this good faith and manipulate this system to meet their own disingenuous needs, but that scenario exists in any system that humans operate in.

In most situations, people do not intentionally misrepresent their IP legal rights. Sometimes the error is just a matter of administrative oversight. The person who knows there is a twenty-year term limit to a patent is usually not the same person working in production and this person is usually separate from those in the marketing department. Each of these individuals is thinking only of their own job responsibilities and not necessarily whether the IP legal rights notices are current. Instead, it will most likely be your competitor or a disgruntled employee who notices and takes action on your mislabeling.

Product Life and IP Longevity

There are times when patent rights last much longer than the product cycle. You may have patent rights for twenty years but only sell the product for eight. This is the reality of product development in a dynamic market. One benefit of U.S. *plant patents* is that there aren't maintenance fees due after the patent grants. Even if the plant is no longer in production, at least you are not bearing the costs of government maintenance fees. However, for sexually-reproduced plants protected by *utility patents*, there are maintenance fees and those fees increase significantly as the patent rights age. In the U.S., for example, at the time of this writing, the third maintenance fee, due at eleven-and-a-half years of patent rights, is up to $7,400 USD.

Returning back to the importance of administration tasks associated with IP portfolio management, if you automatically pay all your patent maintenance fees without first confirming the plants covered by that particular patent are still on the market, then *you are wasting thousands of dollars unnecessarily*. If your plants are protected by PBRs, most countries have annual maintenance fees for each country where you have rights. It is not uncommon to pay $20,000 USD in *annual* maintenance fees for a plant that has multiple PBRs. In such situations, you must decide if the sales activity of the protected asset is generating enough revenue to justify the $20,000 in maintenance fees. Often, it isn't. The financial health of your company will depend on correlated data regarding patent rights and product sales.

In other situations, if you have a utility patent, or multiple legal *claims* articulated in the patent, then throughout the patent grant, circumstances may dictate that some of these claims are no longer valid, and labeling a plant as "patented" may no longer be the case. Continuing to use such labeling may constitute false marking.

In contrast, trademarks technically do not have an expiration term. They can maintain their legal rights *indefinitely* as long as certain criteria are met. That criteria include paying maintenance-type fees to the government and continuing to use the trademark in commerce in the same manner associated with the same goods or service as what you initially told the U.S. Trademark Office you would use it for.

As we see from Erik's situation, there are a lot of potential negative ramifications lurking behind seemingly mundane administrative details. If you lose in this area, then it is very easy to wipe away any of the profits garnered from your patented plant. If you aren't willing to devote the resources to establish a good IP portfolio management system, then you may, in essence, be spinning your wheels. Unfortunately for Erik, his long, stellar career was marred by both his and his company's neglect to address this administrative detail.

CHAPTER ELEVEN

Delano's Story: Public Access, People & Systems

A final major pitfall in the patenting process of plants is *public access*. For a U.S. patent, it is *not* possible to obtain patent protection for an invention that has been in the "public domain" *for more than a year before filing* the application. The U.S. recently moved to awarding patent rights on a "first inventor to file" basis rather than a "first to invent" basis. A "first to file" system has been in place in many other countries around the world for a long time. Therefore, providing access to your plant material *before* your patent application is filed is risky. The following is a true story between Delano Family Farms in California and the United States Department of Agriculture (USDA).[20] This story demonstrates many

20 No attempt is made here to interpret the facts presented in the public records and published opinions pertaining to this case. This case is represented for illustrative, educational purposes only. Facts and parties involved have been consolidated for storytelling purposes. Case No. 1:07-cv-01610-LJO-JLT.

of the potential challenges between developing a plant line and using the patent system.

Days of Yesteryear – The Lone Inventor

In the horticulture industry, patents are specifically about people, partnerships, and plants. Contrary to thinking of people and partnerships, our historic impression of an inventor is an isolated, disheveled tinkerer building contraptions and causing minor explosions in his garage at the expense of showing up to meetings on time or even maintaining personal hygiene.

In the plant world, there are a few such lone rangers left. Scarce are those combing the earth to find the most remarkable plant specimen and successfully propagating it in a private 4 x 8 foot greenhouse. After many years of toil; all the stars align, the selected plant proves hardy, phenotypically stable, has marketable usefulness and/or aesthetic, and even grows in abundance – a perfect plant for commercialization. The plant breeder becomes rich licensing their plant and has enough money to travel to even further reaches of the globe at the mere possibility that a plant exists which he can further develop.

These are the fairy-tales that cultivate the romantic mystique of being an inventor in the world of horticulture. I am not here to dispel our notions, but just as we've had to come to terms with globalization impacting our markets, the landscape for invention in the horticulture industry reveals that, in our modern times, the lone inventor very rarely does it alone.

Patents and PBRs Are About People

In the heartland of the sunny state of California, two table grapes were creating quite a scandal. 'Scarlet Royal' and 'Autumn King', like petulant royals, maintained their innocent look in front of the cameras; meanwhile their behavior behind the scenes was called into question. It may be tempting to label their behavior as outright illegal. Or, perhaps they were merely caught between two worlds as the culture transitioned from an open and free cultivar-sharing philosophy to a patent-protected market evolving to maintain its competitiveness.

It was the end of the 1990s when the USDA suspected that they had two table grapes worth their weight in gold. In order to test their theory, they needed to get other people involved. The grape vines were sent to a trustworthy professor at Fresno State. The professor would test the consistency of the grapes' characteristics. Were they distinct? Were they uniform? Were they stable? These three criteria, commonly referred to as DUS (Distinct, Uniform, Stable), are the international benchmark for a strong, new variety and a candidate for a new patent or PBR.

The grapes were delivered to the professor; however the USDA and professor didn't sign a written agreement detailing the testing arrangement between them. They figured they didn't need a formal agreement. After all, they had been working together for a long time and both parties understood the professional courtesies of the industry. There certainly was no need to make anyone feel awkward in order to hammer out details that were already assumed by both parties. A handshake would suffice.

Scarlet Royal and Autumn King performed so magnificently in testing plots at the university that the USDA decided it was time to introduce the grapes to a group of local farmers. They also felt it wasn't sufficient to merely tell the farmers about the grapes.[21] The USDA knew that the farmers would be most convinced of how amazing these two new grape varieties were if they could actually see and touch the grapes themselves.

Mr. Jim Ludy, a local farmer attending the presentation, was very impressed. Later he was talking with one of his friends who worked at the USDA. This friend wasn't directly involved with Scarlet Royal and

21 A little poetic license has been taken by the author to improve storytelling and to support what commonly happens in the industry. No one, including the USDA was actually interviewed to inquire on their actual emotions and reasons taken behind the actions. It is mere conjecture on the part of the author to demonstrate the challenges with patenting and managing plant development programs.

Autumn King, but he and Mr. Ludy were buddies and he knew that if the grapes were tested in another plot and their DUS traits were verified again, that could only improve the market value for the grapes. Even though he didn't think it needed to be said, the USDA employee emphasized to Mr. Ludy that the grapes were to be planted *only* on Ludy's land and that Mr. Ludy should keep the testing results private. Certainly, Mr. Ludy was in agreement, after all, this is how it was done. It had been done this way for so long, that there really was no need to even say it aloud. Naturally, the USDA employee and Mr. Ludy didn't sign a written testing and confidentiality agreement.

After several years of testing, the USDA filed for patent applications. In 2005 they were finally rewarded with two granted plant patents. In order to recoup the financial investment in research and development and intellectual property preparation fees; the USDA began to collect fee premiums from the farmers who licensed the grapes and sold the grapes for commercial products. This arrangement functioned smoothly for about a year before the USDA found itself in court.

In 2007 the Delano Family Farms, et al. claimed that the patents were invalid because the grapes were available to the public for more than a year before filing the patent application. Since the patents were invalid, they argued, they didn't need to pay a royalty for the table grapes they sold and there shouldn't be a restriction on propagating the vines. In addition,

Delano revealed that the grape vines had already been growing on Mr. Ludy's *brother's* farm more than one year before the patent applications were filed – and, therefore, they were publicly available. How the grapes arrived at the brother's farm is still being determined and at the time of this writing, the case is still pending.

The situations in this case are illustrative of how to design your own plant breeding programs to avoid the many potential pitfalls of Scarlet Royal and Autumn King – and to not have your patents go *POOF!*

The Problems with Material Sharing

As previously mentioned, entrenched in the plant breeding culture is the concept of sharing. Experts readily share their hard-earned knowledge with others. Seeds and cuttings are freely and generously given to neighbors and colleagues in abundance. The recipients, in turn, give those seeds and cuttings to their peers. This habit of sharing is not only attributed to a generous spirit; sharing has its practical purposes, too. Plants respond differently depending on their growing environments. It takes many years and, ideally, varying growing conditions to test the plant's characteristics. In fact, the data presented in plant breeder's rights and patent applications are considered more reliable if the plant is grown in different plots, in different locations, and over several generations and years.

The problems begin when the conditions surrounding the testing of plants to determine if they are viable candidates for patents, conflicts with the U.S. Patent Office's fatal ruling that *an invention cannot be publicly available for more than a year before filing.* The story of Scarlet Royal and Autumn King perfectly demonstrates the intersection of variety testing and patent law.

If you've been reading this book sequentially, then you will not be surprised by my mantra that in having good systems in place, you can avoid this quagmire altogether and avert the court system. No matter how reluctant you may be to have written contracts and spreadsheets to keep track of your plant development, and resigning yourself to filling out additional paperwork, these tools are the rules of the patent game. *Generous plant material sharing during the development phase will likely conflict with patent protection rules.* If you don't employ these tools and you still file for intellectual property protection, then you may be better off shredding your money and turning it into compost.

Delano, et al. claimed that the patents were invalid because the grapes were available to the public for more than a year before filing the patent application, which automatically bars a patent application from being granted. *Patent invalidity* is a legal term asserting the existence of an error in the written portion of the patent itself or the inventive process leading up to the patent grant. These errors, if demonstrated true, disregard or limit legal rights claimed in the patent. If the IP court and Patent Office "gods-that-be" (i.e. the patent examining board) determine that a patent is invalid, *all aspects* of an invention become publicly available. In practical terms, this means that the owner of the now-invalid-patent cannot restrict the use of the invention, nor demand royalties for use.

The Drop Dead Rule

The term "drop dead" is as bad as it sounds. There are certain deadlines in a patent application's life that must be met or the application is rejected, invalidated, or otherwise abandoned. Generally, there is a lot of flexibility in the patenting process. Plenty of opportunities exist to recover from mistakes and missed deadlines, as long as you are willing to spend the money. However, the U.S. Patent Office's one-year filing from public access to the invention filing date rule is *not* one of them.

The intent behind the one-year law is to encourage a more rapid progress in innovation. If the Patent Office allows for exclusive rights over an invention for twenty years from the date of filing, then they do not want the inventor to sit on the invention for years at a time before filing a patent application. The Patent

Office's intent behind this rule is that technology will advance at a healthy pace *if* deadlines exist for filing patent applications. In other words, the Patent Office understands human nature and understands that most of us finally finish something only when there is a looming deadline.

Conversely, the U.S. Patent Office also recognizes that it is economical to commercially test a market before incurring the expense of filing a patent application. So, they grant a one-year grace period between public availability and the patent application filing date.

Generally speaking, I do not find U.S. patent law synchronous with normal plant development businesses because it takes many years to develop each plant product. This is longer than other industries that often use the patent system, such as the electronics industry. Only a small percentage of the patent applications filed each year are for plants. Out of the 576,763 applications filed in the U.S. in 2012, only 1,149 (or less than one-tenth of 1%) were plant patent applications.[22] This figure does not include the utility applications for commodities like soybean and corn, of which there are many more applications. However, at least from a plant patent perspective, you can see that this is a niche area of the patent process.

There are few IP professionals who specialize in this sector of intellectual property. There also aren't many

22 United States Patent and Trademark Office: Patent Technology Monitoring Team (PTMT). (N.D.) U.S. Patent Statistics Chart: Calendar Years 1963-2013. Retrieved on April 12, 2014 from http://www.uspto.gov/web/offices/ac/ido/oeip/taf/us_stat.htm Note: The data did not segregate how many utility patents were filed specifically for plants.

plant-specific court cases that Patent Offices and courts can use as a framework for their decision-making. All of these factors combine for my motivation in writing this book. Understanding the system and creating a strategy that matches your commitment and resources can avoid most of the pitfalls here; all of which can help you use IP for a competitive advantage and profit.

What Happens When One Year Isn't Enough?

Plant-specific IP laws went into effect in the U.S. in the 1930s. These new plant patent laws provided an excellent tool for protecting plants that are propagated asexually. Then, in 1970, the U.S. adopted a version of the international UPOV Convention treaty,[23] which offers a plant variety protection (PVP) system for sexually-reproduced plants. Finally, in the 1980s, the U.S. began to allow for patent protection for sexually-reproduced plants through a *utility patent*. Each of these systems has a one-year public access filing date requirement.[24] One obvious deficiency from a plant breeding perspective in

23 UPOV stands for International Union for the Protection of New Varieties of Plants and is a global treaty providing streamlined guidelines for obtaining plant breeder's rights protections in the countries that adopt the treaty.
24 UPOV's filing dates relate to the date of public access or offer for sale within a particular country. Then, in some cases, you can have up to four years to file in other countries. There are many variations to this last statement that are beyond the scope of this book. If you need specific information on what your "drop dead" filing dates are, it is best to consult with your IP professional who has expertise with PBRs.

patent protection requirements is in this one-year grace period. This requirement is a challenge for many plant breeders, including those working for the USDA in the Delano case.

In other industries, it is common when developing a prototype, to build it at scale for the commercial market, offer the product for sale, and determine whether it is viable. Like plant breeders who use growers to produce the plants at commercial scale, other industries often rely on outside partnerships to help them bring the technology to market. These relationships have defined roles and usually have strong *confidentiality contracts* to control and preserve the secrecy of the technology.

Also, it is not uncommon for companies in other technical industries to file multiple patent applications on a technology as the technology evolves. Filing multiple patent applications in the horticulture industry is, in most cases, impractical or simply not cost effective. With the exception of a utility patent, which can protect *portions* of a plant (such as alleles identified for expressing distinctly characteristic qualities) patents for plants are for the entire genome of a plant. Further, in most cases, an individual plant is not sold with a high enough royalty at a high enough volume to warrant filing more than one patent application related to that particular plant product because the cost of filing multiple patent applications is prohibitive. *Compounding the situation is the fact that one year is usually insufficient to develop a plant and to test its market viability.*

So, what should plant breeders do? They must rely on maintaining secrecy and tightly control access to their plant for the entire plant development and production stages until an application is filed. If a plant breeder wants to use patent laws to provide a higher royalty and commercial advantage for a particular plant, then the plant breeder must also rely on other formal, legal conventions, such as written agreements and careful record keeping.

Loose Lips Sink Ships: Public Access & Confidentiality

The Delano story demonstrates how patents are about people and partnerships. In this scenario, plant material is being sent from the USDA (who has many employees) to a university (who has many employees). Even if the plant material is under the direction of a competent professor, well-versed in cultivar testing and confidentiality protocol, there are many opportunities for the plant material to become publicly available.

Perhaps an over-zealous summer intern or student decides to take a few cuttings of the plant and share them with others, as we saw in Raymond's story. These situations demonstrate the key role of people and partnerships as it relates to patents.

So, What *Exactly* Is Public Access?

That is the million-dollar question, isn't it? Public access can include a public demonstration of a plant, or perhaps a gift given to a public garden, especially when the public can actually put their hands on the plant long enough to obtain a clipping. It can include a printed publication, especially when coupled with any of these other scenarios. Public access also includes a sale or an offer for sale. Then, further down our Alice in Wonderland rabbit hole, one can ask: "What constitutes an *offer for sale*?" Growers' contracts, as we saw in Yoshi's story, can inadvertently trigger an offer for sale depending on the language in the contract. A sale is confirmed even when there is only one buyer purchasing as little as one plant for just one dollar. This sale still counts towards your "one-year" drop dead due date even if that person changes their mind and never gives you that dollar. It was still a formal *offer* for sale.

These are just some of the situations that may qualify as public access and trigger a *POOF!* of your patent potential. The Patent Office enjoys offering a number of colorful examples that will frighten anyone into maintaining secrecy, so much so, you may even try to grow your plants under a tarp – which of course probably will not help your plants very much.

The Insufficient Handshake

There are many problems with non-written agreements and the desire to maintain patent rights. In fact, if you don't want your IP portfolio to go *POOF!*, I can safely say that non-written agreements are incompatible with obtaining solid patent rights.

One reason written agreements are generally avoided is because they involve tedious paperwork. Paperwork is generally something any average person wants to avoid, especially by the outdoor-oriented personalities of the people who are attracted to the horticulture industry.

I can appreciate that in the horticulture industry in particular, we all want to believe that a person is as good as their word and a handshake is binding. The people I meet and work with are "good people" and I expect we all have similar assumptions. It is uncomfortable to lay out your assumptions and expectations of another person in writing, just as it is uncomfortable to plan your death, in the form of a will, while you are still living. It is uncomfortable to iron out details of a premarital agreement when you are madly in love. And, it is much more comfortable to believe that your business partner will always want to work *with you* instead of *against you,* so that you both can make the most money. The "hope" that everything will work out well in the end, and that someone is going to stay true to their word, generally works until either: 1) the product is making a lot of money, 2) there is a perceived inequity in the partnership, or 3) you are faced with legal risks or other obstacles. Remember that formalizing a written contract can feel very uncomfortable for both parties and can lead to misunderstandings. In the worst case situations, these scenarios deteriorate a long-term relationship.

In spite of this awkwardness, a handshake is not the solution because the details behind a handshake become murky over time. It is difficult to assert legal responsibility to any person involved in a verbal agreement. Signing a contract, even if it makes someone feel a slight discomfort, suggests that this person should feel that same discomfort when faced with opportunities to violate what was agreed to. The farmers who accepted the vines for testing presumably agreed, or certainly understood that the vines were to be under their control and not shared with anyone. However, somewhere along the way, the farmer's brother ended up growing the vines on his land. Surely this is not what the USDA would have wanted because they were still in the testing phase of their cultivar development and had not filed a patent application yet.

Play It Close to the Chest

The courts have a conflicted history on deciding whether or not an invention owner is responsible for complete control to the access of their invention or whether they can have any legal recourse if someone steals their invention. When you don't know which way the wind will blow and there isn't clear direction in the law as to whether your systems and behavior will be legally supported, my suggestion is to make sure your systems are tight so that you never have to test the varying legal weather.

Were the vines in the Delano case stolen? In my opinion, probably not. Perhaps they were shared in the same good spirit that many farmers share their plants. Was the USDA employee violating protocol? Who knows? Perhaps a written protocol that was developed by the legal team was overlooked, but the history and the culture of any plant breeding organization may imply that what the USDA employee did was not out of line. The main point is that the more people involved in the process, the more opportunities there are for the inventors and plant material owners to lose control over the material. That being said, what can you do about it?

How well-organized are your plant development records? Considering it took eight years for the Delano case to emerge, could you account for all the development decisions of your plant portfolio? Could you fully account for all of those who had access to your plant material for eight years? Have you recorded the dates when plant material was released for testing or propagation? Who keeps these records up to date? Are there any gaps in the data, say, when one employee left and another took over this responsibility?

At the time of developing a plant, investing this much resource into the administrative duties may feel like overkill. There is an expression: *He who has the most documentation wins.* If you find yourself working with a litigation attorney, you certainly want to find yourself with the advantage of having clear and consistent records.

CHAPTER TWELVE
Conclusion

Making mistakes through trial and error in the complicated world of intellectual property can be far less forgiving and have far more severe and permanent consequences than other aspects of business. Therefore, it is critical that things be done right the *first time*. With IP, it is often the things you do *not* do, do *not* engage in, do *not* pursue, and do *not* forget about that are just as important as the things that you do engage in that guarantees your portfolio's success. *Poof!* is intended to facilitate successful IP management by showing business owners and IP managers how to learn from OPM – Other People's Mistakes. You don't have to make all of these mistakes yourself, and you don't have to reinvent the wheel. Each story of *POOF!* represents one or more valuable lessons about mismanaging or misunderstanding intellectual property portfolio management.

In Susan's story, we learned first that trademark protection only protects you from other parties using your plant's *name*. It does not prohibit others from propagating your *exact* plant and selling it under a *different* name. Susan also found out the hard way that unless you specify otherwise, a trademark only applies to a certain kind of goods or services. It doesn't offer across the board protection for all products across any sector. Someone with the same or similar trademark as yours can sell products in a different class without infringing on your trademark. Susan also discovered that trying to "save on IP costs" by filing intellectual property instruments on her own without the guidance of an IP professional can really cause long-term harm to your competitive edge in business. Lastly, Susan discovered how and when to transition her trademark into a full-fledged brand in order to generate more revenue through licensing opportunities, and we examined the role of brand promises and partnerships to help ensure profitable brand extensions.

Jim Clark and Charbucks Coffee Blend helped us understand the potential ramifications of using marketing names that share similar elements with famous, trademarked brands. We explored the challenges associated with letting our emotions distort our perception of risk and we saw that even if you win in court, it doesn't mean you have actually won anything of value. Seth, our plumber who deliberately "borrowed" on the public recognition of the Superman logo, helped us further explore the relationships among trademarks, branding, and marketing. We came to understand that when a trademark owner asserts their rights against a potential infringer, they aren't doing so because they are a bully, but rather because they are legally expected to do so, unless they wish to risk losing some or all of their trademark's rights and value. We concluded our discussion on trademarks by specifically reviewing what kind of trademarks will be outright rejected by the Trademark Office and, more importantly, how to submit a trademark that stands a good chance of being accepted first time around – potentially saving applicants thousands of dollars in expenses, wasted time, and frustration.

Margaret then introduced us to the challenges of operating in a global market and how developing a strategy without carefully-monitored controls can cause costs to run amuck, even before you actually file any applications. Some simple, but helpful tools were offered to try and minimize the associated fact-finding costs, especially as it related to seeking international patent and trademark information. Yoshi's experience with filing applications in Canada demonstrated to us how important it is for everyone on the IP application team to have access to comprehensive and contextually-relevant information so that critical filing dates aren't missed, thereby jeopardizing an asset in one's portfolio. International plant IP poses so many caveats, it is difficult to anticipate how specific circumstances intersect with the facts. Perhaps most importantly, Margaret's and Yoshi's stories opened our eyes to just how complicated and cumbersome international filings can be.

Erik's story demonstrated another aspect of the importance of keeping up-to-date on administrative IP management responsibilities like patent expiration dates. We saw how even an innocent oversight could be construed by some as much more sinister, and perceived as deliberately deceptive, tarnishing one's hard-earned reputation in the industry.

Walter and his trees helped us understand how important it is to not put the cart ahead of the horse – that is, to conduct proper sequencing. Though it is common to discover new plant inventions, it is critical that we confirm *all* the circumstances of the discovery and the cultivated area *before* we assume that a new plant is patentable and invest limited, valuable resources, only to be flatly denied a patent.

Raymond and Jaime introduced several complications of patenting plants, namely that patenting is a separate part of your business and that you may not want to engage in it merely as a "side project." We looked at the differences between ownership and inventors and even introduced how complicated, time and resource consuming, and energy draining it can be to determine who the inventors are when conflict arises. We discussed the importance of recognizing how powerful the notion of perceived ownership can be among employees, and how best to address it. Finally, Raymond and Jaime learned about the consequences of not taking the time to carefully read or remember the signing of assignment contracts.

Delano's story illustrated the challenges of developing plants while maintaining confidentiality and tightly limiting access to plant material *before* any patent applications are filed. We also witnessed that the Delano story spans a transition in the horticulture industry where, if you want to financially benefit from the patent system, then you need to be willing to play by their rules and transition to a paperwork and contract-driven environment. Lastly, we provided a risk assessment tool so that you can better evaluate what types of intellectual property might work best for you.

The people in these stories are remarkable, talented people conducting business by creating valuable, yet everyday products – our morning coffee, the grapes we have with breakfast, the trees and ornamentals lending grace to our personal gardens and our dinner vegetables. These items contribute to the magic of our everyday lives. Intellectual property, like any good magic trick, at first appears wondrous and astounding. But as with all magic tricks and illusions, they always have a simple, logical explanation, and are executed through practiced, repetitive, carefully planned systems and a strong attentiveness to detail. By following these systems, you can prevent your potentially-lucrative plant assets from going *POOF!*

About the Author

Allison's Story

As an Intellectual Property professional for more than a decade, Allison Frederick has successfully managed clients' domestic and international trademark, patent, and plant breeder's rights applications and portfolios, including those of Fortune 500 companies. Allison was selected to catalogue the historical intellectual property portfolio for Western Union from 1851-2005 at the Smithsonian Institute archives in Washington, D.C.

Allison is the owner of GReenhouse Intellectual Property Systems (GRIPS, LLC), serving the intellectual property management needs for clients all over the United States and abroad. With connections to intellectual property specialists all over the world, she is a powerful resource for horticulture and agriculture organizations of any size. She is available for consultation (both on-site and remote), speaking engagements for the horticulture industry, and for active management of your organization's intellectual property portfolio. Allison can be contacted through her www.greenhouseIPS.com website or by emailing Allison at allison@greenhouseIPS.com.

Allison is presently completing her Masters' Degree in Sustainability and Environmental Management at Harvard University Extension School. She is also a member of the Denver Women's Press Club, which has been in existence since 1898. In 2014, Allison was appointed by the Governor of Colorado to a small business advisory board.

CPSIA information can be obtained at www.ICGtesting.com
Printed in the USA
LVOW06*2321031215

465286LV00006B/36/P

9 780991 564637